Nuclear Infrastructure Protection and Homeland Security

FRANK R. SPELLMAN AND
MELISSA L. STOUDT

GOVERMENT INSTITUTES
An imprint of
ROWMAN & LITTLEFIELD PUBLISHERS, INC.
Lanham • Toronto • Plymouth, UK
2011

**Government
Institutes**

Published by Government Institutes
An imprint of The Scarecrow Press, Inc.
A wholly owned subsidary of The Rowman & Littlefield Publishing Group, Inc.
4501 Forbes Boulevard, Suite 200, Lanham, Maryland 20706
http://www.govinstpress.com

Estover Road, Plymouth PL6 7PY, United Kingdom

British Library Cataloguing in Publication Information Available

Library of Congress Cataloging-in-Publication Data

Spellman, Frank R.
  Nuclear infrastructure protection and homeland security / Frank R. Spellman and Melissa L.
Stoudt.
    p. cm.
  Includes bibliographical references and index.
  ISBN 978-1-60590-713-0 (pbk. : alk. paper) — ISBN 978-1-60590-714-7 (electronic)
  1. Nuclear facilities—Security measures—United States. I. Stoudt, Melissa L. II. Title.
TK9152.1645.S64 2011
363.325'9629134355—dc22

                                                                                          2010043780

∞™ The paper used in this publication meets the minimum requirements of
American National Standard for Information Sciences—Permanence of Paper for
Printed Library Materials, ANSI/NISO Z39.48-1992.

Printed in the United States of America

Contents

Preface

The fifth of the new Government Institutes series on critical infrastructure and homeland security, *Nuclear Infrastructure Protection and Homeland Security* is a reference source that serves U.S. energy production businesses and managers who want quick answers to complicated questions and helps nuclear reactor, materials and waste sector employers, and employees handle security threats they must be prepared to meet on a daily basis. In the post–September 11 world, the possibility of nuclear energy–infrastructure terrorism—the malicious use of weapons to cause devastating damage to the nuclear energy industrial sector along with its cascading, radiating effects—is very real.

Prior to 9/11, as part of various security consulting assignments, we made several collateral assessments of energy infrastructure site security at facilities primarily involved in energy production (also applicable to the U.S. nuclear power industry), storage, and/or usage. Site audits were comprehensive in nature and based primarily on the requirements of OSHA's Process Safety Management Standard (PSM) 29 CFR 1910.119—which outlines requirements for preventing or minimizing the consequences of catastrophic releases of toxic, reactive, flammable, or explosive chemicals and U.S. EPA's Risk Management Program (RMP). From interviews with site managers and security staff personnel, we found the following concerns:

On the one hand

- U.S. energy infrastructure sites were well aware of their vulnerabilities and are making significant efforts to increase planning and preparedness.
- Cooperation through industry groups has resulted in substantial information sharing for effective and best practices across the sector.
- Many site owners and operators have extensive experience with infrastructure protection.

- Many site owners and operators have recently focused their attention on cyber security.

On the other hand

- Overall, security at U.S. energy infrastructure locations (excluding nuclear reactor, materials, and waste sector sites) ranged from fair to very poor.
- Excluding nuclear power infrastructure, energy site security managers were very pessimistic about their ability to deter sabotage by in-house employees, yet none of them had implemented simple background checks for key employees such as energy production operators (all nuclear power operating personnel are thoroughly screened).
- None of the corporate security staff had been trained to identify combinations of common chemicals at their sites that could be used as improvised explosives and incendiaries.

In one confidential client's report, we noted that

> among the "soft targets" identified as potential terrorist sites/targets were hydroelectric dams, electrical distribution power stations and transmission towers, petroleum refineries, fuel transport and end-user distribution sites, the coal industry and its associated ancillaries, the nuclear power industry, the renewable energy industry [wind power (especially wind farms), solar power, and alternative fuel manufacturers and distributors], silviculture (forest) product production (firewood as a source of fuel), along with compressed gases in tanks, pipelines, and pumping stations; and pesticide manufacturing and supply distributors.

Typically, the energy industry comprises (Energy Industry 2009)

- the petroleum industry, including oil companies, petroleum refiners, fuel transport, and end-user sales at gas stations.
- the gas industry, including natural gas extraction, and coal gas manufacture, as well as distribution and sales.
- the electrical power industry, including electricity generation, electric power distribution, and sales.
- the coal industry.
- the nuclear power industry.
- the renewable energy industry, comprising alternative energy and sustainable energy companies, including those involved in hydroelectric power, wind power, and solar power generation, and the manufacture, distribution, and sale of alternative fuels.

- traditional energy industry based on the collection and distribution of firewood, the use of which, for cooking and heating, is particularly common in poorer countries.

In this book, however, we focus on the three interrelated nuclear energy infrastructure segments: nuclear reactors, radioactive materials, and nuclear waste.

This book (and the other twelve forthcoming volumes of the critical infrastructure series) was written as a result of 9/11 to address these concerns. *Nuclear Infrastructure Protection and Homeland Security*, in particular, was fashioned to address the critical needs of nuclear energy production managers, nuclear energy conversion to electrical power and transmission and distribution managers, nuclear energy infrastructure engineers, nuclear design engineers, nuclear process managers at any level of energy production, engineers, environmental professionals, security professionals (physical and cyber security), industrial hygienists, students—and anyone with a general interest in the security of nuclear energy infrastructure systems. It is important to point out that peak oil—the point in time when the maximum rate of global petroleum extraction is reached, after which the rate of production enters terminal decline—is rapidly approaching. There are many predictions as to when this will occur, with many experts believing that the only way to determine it will be in retrospect, at which time it will be too late to be unprepared. Simply, nuclear power is important now but soon will be critical to maintaining our way of life. Acknowledging that nuclear energy is soon going to be a necessity, it is important to point out that our nuclear energy infrastructure (as is the case with the other sixteen critical infrastructures) cannot be made absolutely immune to all possible intrusions/attacks. It takes a concerted, well-thought-out effort to incorporate security upgrades in the retrofitting of existing systems and careful security planning for all nuclear energy infrastructure components. These upgrades or design features need to address issues of monitoring, response, critical infrastructure redundancy, and recovery to minimize risk to the facility/infrastructure.

Nuclear Infrastructure Production and Homeland Security presents commonsense methodologies in a straightforward, almost blunt manner. Why so blunt? In this era, when dealing with security of workers, family members, citizens, and society in general, the right to live in a free and safe environment is a reasonable demand.

This text is accessible to those who have no experience with nuclear energy infrastructure and/or homeland security. If you work through it systematically, you will gain an understanding of the challenge of domestic preparedness—that is, an immediate need for a heightened state of awareness of the present threat facing the nuclear energy infrastructure industrial sector as a potential terrorist target. Moreover, you will gain knowledge of security principles and measures that can be implemented, not only adding a critical component to your professional knowledge but also giving you the tools needed to combat terrorism in the homeland—our homeland.

Frank R. Spellman
Norfolk, Virginia
Melissa L. Stoudt
Burnt Hills, New York

REFERENCES AND RECOMMENDED READING

Energy industry. 2009. http://en.wikipedia.org/wiki/Energy_Industry (accessed April 14, 2009).

Prologue
A Trip to Town

Day rose heavy and hot, but the wind whispered in the field beyond the sod house as if murmuring delightful secrets to itself. A light breeze entered open windows and gently touched those asleep inside. A finger of warmth, laden with the rich, sweet odor of earth, lightly touched Alena's cheek—rousing her this morning as it had often in her nine years of life. On most days, Alena would lie on her straw mat and daydream, languishing in the glory of waking to another day on Mother Earth. But nothing was normal on this morning. This day was different—full of surprises and excitement. Alena and her mother Zoya were setting out on adventure today—and Alena couldn't wait.

As she stood at the foot of her makeshift bed, Alena swiftly slipped her best brocaded silk, shapeless sarafan over her head and in place, covering her slender shape, and let the fall of cloth hang to her feet. She pulled her straight black hair tight in a knot at the back of her neck before she placed the round sarafan-matching kokoshnik on top of her head.

While Alena dressed, Zoya performed the same ritual in her small room, next to Alena's. Zoya was excited about the day's events, too—She knew Alena was thrilled, and she was delighted in her daughter's pleasure and excitement. Zoya chuckled to herself as she remembered the many times over the past few years that Alena had begged to be included, to be taken to Pripyat, the small town nearest their modest home. Zoya understood Alena's excitement. Going into the town, taking it all in—shops, stores, the market—thrilled Zoya, too.

As they stepped out of the sod house and onto the dirt road, the scented breeze that had touched Alena's cheek earlier greeted them. They walked together, hand in hand, toward town, twenty-seven kilometers to the south. They knew they had an all-day and half-the-night walk ahead of them before they reached a relative's home in town—but they did not care; at that moment, they felt they could walk forever.

Alena bubbled with anticipation, but she held it in, presenting the calm, serene face expected of her. Even so, every nerve in her young body reverberated with excitement.

As they walked along the road, Alena, fascinated by everything she saw, took in everything they passed in this extension of her small world. People and cattle everywhere—she had never seen so many of either! Her world had grown, suddenly—and it felt good to be alive.

After their steady all-day walk it was now early in the morning, and in the moonlight Alena could see tall buildings of the huge industrial facility they were passing and the lights of the city on the other side. How big and imposing they were—and so many towers and buildings! Outside the industrial facility, in sections they passed, some of the roads were actually paved. Alena had never seen paved roads or streets. This trip to town was her first city experience, and though only slightly lighted in moonlight she was enthralled by all the strange and wonderful sights. As they walked along the street leading to the town, Alena was overawed by the tall buildings and warehouses. "What could they all be used for?" she wondered. Some of them had signboards above their doors. She saw a huge sign above the gate to the industrial complex, but little good that did for Alena—she couldn't read.

The light, following breeze had escorted Alena and Zoya since they left home, and it was still with them as they turned toward the city. Alena could see more city lights and the ghosts of buildings ahead, and her eyes grew wider with excitement.

Suddenly, behind them there was a huge explosion with an accompanying blast of strong wind that picked them up, carried them a few feet, and set their shattered bodies down abruptly. And with one breath of that sweet air (was it the same sweet air that had touched her into waking only twelve hours earlier?), Alena began vomiting profusely. She rolled over on the pavement; her body felt as though it were baking—and it was. Her mother was also fallen, clutching her abdomen in pain. But Alena didn't have time to realize what was happening. She couldn't think. She couldn't do anything—except die—and she did.

Alena, Zoya, and several others died almost instantly.

Those who died that April 26, 1986, day never knew what killed them. However, the several thousands of others who died soon after did know what killed them.

The inhabitants who lived near the town, near the industrial complex, who worked inside the industrial complex, near the release point, knew only death and killing sickness in the coming days.

Those who survived the first week were later told that a lethal dose of radiation had killed their families, their friends, their neighbors, their acquaintances. They were killed by an accidental explosion that today is infamous in the journals of hazardous materials incidents. Today, this incident is studied by everyone who has anything to do with radiation and the securing of nuclear facilities from accidents, vandalism, and/or terrorist acts. We know it as Chernobyl.

1

Introduction

The Nuclear Energy Institute (NEI 2009) points out that U.S. nuclear power plants were safe and secure before September 11, 2001, and they are even more secure today. Moreover, nuclear power plants are the most secure industrial facilities in the nation. The industry incorporates physical security measures, plant surveillance and access control systems, and SCADA security and cyber security, and it coordinates with federal, state, and local law enforcement authorities to keep nuclear plants safe and secure. Here's what the Nuclear Regulatory Commission (NRC) says about nuclear power plant security:

> The NRC remains convinced that nuclear power plants are among the most heavily protected civilian facilities in the Unites States. Our vulnerability studies confirm that the likelihood of damaging the reactor core and releasing radioactivity that could affect public health and safety is lower. All nuclear power plants have been required to enhance their safety security, and emergency preparedness. Given these enhancements, the potential radiological consequences to the public of an aircraft attack are low (Diaz 2004).

U.S. NUCLEAR INFRASTRUCTURE: A TOUGH NUT TO CRACK*

The many components to U.S energy infrastructure, including nuclear power generating plants, waste transportation, and waste holding and storage sites, are everywhere, in every direction, in almost endless combinations, interrelationships, and quantities. In making this simple observation, it matters little whether potential terrorists are of foreign or domestic origin. The fact is they only need to look around or surf the Internet to locate the components of the U.S. energy sector. In doing this, terrorists must be in a state of utter fascination, with many seemingly good or tempting targets all around them.

---

* Information contained in this section is adapted from *America's Energy Infrastructure: A Comprehensive Delivery System*. Retrieved April 16, 2009, from www.netl.doe.gov/publication/press/2001; U.S.NRC (2008) *Security Spotlight*.

Let's take a look at what anyone, including terrorists, can find out about our energy infrastructure.

- In the somewhat desolate Columbia-Snake River basin of Washington and Idaho states and surrounding the Hanford Site (a.k.a. Hanford Works, Hanford Project, Hanford Nuclear Reservation), with its nine decommissioned weapons production nuclear reactors and 53 million gal (204,000 m³) of high-level radioactive waste, are several hydroelectric dams, strung like beads. These dams consist of Grand Coulee, Chief Joseph, Wells, Rocky Reach, Rock Island, Wanapum, Priest Rapids, and Ice Harbor. Going westward along the Columbia are McNary, John Day, The Dalles, Bonneville dams. To the east along the Snake River are Lower Monumental, Little Goose, Lower Granite; finally, to the south along the Snake River bordering Idaho-Oregon, Hells Canyon, Oxbow, and Brownlee dams.
- At the present time, 104 operable nuclear power plants are located in many states in the U.S. The power plants are listed by group depending on the region of the U.S. in which they are located. Region IV is the largest geographical region with the smallest number of power plants, twenty plants. The region includes the states of North Dakota, South Dakota, Nebraska, Missouri, Arkansas, Louisiana, and all the states west of these. Region III includes the states of Minnesota, Wisconsin, Michigan, Iowa, Illinois, Indiana, and Ohio. This region includes twenty-four nuclear power stations. Region II includes the southeastern states of West Virginia, Kentucky, Virginia, Tennessee, North Carolina, South Carolina, Georgia, Alabama, Mississippi, Florida, and the territories of Puerto Rico and the Virgin Islands. This region includes thirty-four nuclear power stations. Region I includes the New England states of Maine, New Hampshire, Vermont, New York, Massachusetts, Pennsylvania, Connecticut, New Jersey, Delaware, Maryland, and Washington, DC. This region includes twenty-six nuclear power plants.
- Hydroelectric dams and nuclear power plants generate electrical power, of course. However, they are only the best known (the crown jewels) and easily recognized producers of electricity. In the U.S., we also generate electrical power at various plants that use coal, petroleum liquids, petroleum coke, and natural and other gases.
- Producing electrical energy involves more than just its initial production. Along with generation, the transmission and distribution of electrical are also part of the overall production process. The electricity infrastructure includes a North American power grid of long-distance transmission lines that move electricity from region to region, as well as the local distribution lines that carry electricity to homes and businesses.
- Petroleum products are pumped from the ground in domestic oil fields and conveyed through gathering lines to pipelines, which bring it to refineries, where it is

transformed into gasoline, diesel fuel, or heating oil. These products then travel through pipelines and tanker trucks to distribution outlets for purchase by consumers. Finished product pipeline systems involve a complex infrastructure of their own, including compressor stations or plumbing stations and control systems that open and close valves and which produce flow through pipes, often with the use of computer technology.

- Facilities turn raw natural resources into useful energy products.
- Energy infrastructure also includes rail networks, truck lines, and marine transportation.

Again, a terrorist or group of terrorists does not have to look far in the U.S. or on the Internet to find energy and nuclear power plant infrastructure targets. Terrorists also have little trouble in recognizing that energy infrastructure is vulnerable to physical and cyber disruption that could threaten not only its integrity and safety but also the integrity and safety of large regions of the entire country. Moreover, the U.S. transportation and power infrastructures are complex and interdependent. Consequently, any disruption can have extensive consequences.

With specific regard to U.S. nuclear power plants, the Nuclear Regulatory Commission (NRC) requires nuclear power plants to protect against threats. These plants are some of the most fortified civilian facilities in the country. After 9/11, the NRC used its independent regulatory authority to order the nuclear industry to implement new defensive capabilities, more rigorous guard training, and many other security enhancements. In response, the industry has met the increased requirements regardless of cost. The NRC requires that nuclear power plants be both safe and secure. Safety refers to operating the plant in a manner that protects the public and the environment. Security refers to protecting the plant—using people, equipment, and fortifications—from intruders who wish to damage or destroy it in order to harm people or the environment. Upgrading security is an ongoing process (U.S. NRC 2009).

In this book, presented in a generalized format (i.e., ensuring that nuclear power plant security is not compromised in any way), we discuss nuclear power plant infrastructure security and the ongoing process of upgrading security and security training. When you read this book, we feel that you will come to the same conclusions we have, that is, based on our experience and knowledge in all sectors of critical infrastructure, nuclear power security is a tough nut to crack.

When reading this book and learning about many of the comprehensive safety and security measures installed in and practiced by each of the 104 operating nuclear power plants to ensure their own and our protection against terrorist attack, you might be overcome by a warm and fuzzy feeling about the state of nuclear power plant safety and security. However, when basking in that feeling of security provided by physical, cyber, and other security provisions covered in this text, it is important to remember what we pointed out in the other four volumes of this series:

> Never underestimate the time, expense, blood, sweat, and effort a terrorist will expend to compromise the security of any industrial facility.

NUCLEAR EVENT METRICS

Everything that happens, whether a natural event (e.g., hurricane, earthquake, etc.) or some man-made fiasco, is measured in some way. This is also true for nuclear incidents and/or accidents. For instance, in 1990 the International Atomic Energy Agency (IAEA) introduced the International Nuclear Event Scale (INES) as a tool to communicate the safety significance of reported events at nuclear installations or involving nuclear materials to the international technical community.

As shown in figure 1.1, the INES consists of a seven-level event classification system. Events of greater safety significance (levels 4–7) are termed "accidents," events of lesser safety significance (levels 1–3) are termed "incidents," and events of no safety significance (level 0 or below scale) are termed "out-of-scale deviations." The criteria or safety attributes for each event are described below.

- **Level 7**, major accident—Major release: widespread health and environmental effects.
- **Level 6**, serious incident—Significant release; likely to require full implementation of planned countermeasures.
- **Level 5**, accident with off-site risk—Limited release; likely to require partial implementation of planned countermeasures. Severe damage to reactor core/radiological barriers.
- **Level 4**, accident without significant off-site risk—Minor release; public exposure of the order of prescribed limits. Significant damage to reactor core/radiological barriers/fatal exposure of a worker.
- **Level 3**, serious incident—Very small release; public exposure at a fraction of prescribed limits. Severe spread of contamination/acute health effects to a worker. Near accident, no safety layers remaining.
- **Level 2**, incident—Significant spread of contamination/overexposure of a worker. Incidents with significant failures in safety provision.
- **Level 1**, anomaly—Anomaly beyond the authorized operating regime.
- **0 Deviation**—No safety significance.
- **Out-of-scale event**—No safety relevance.

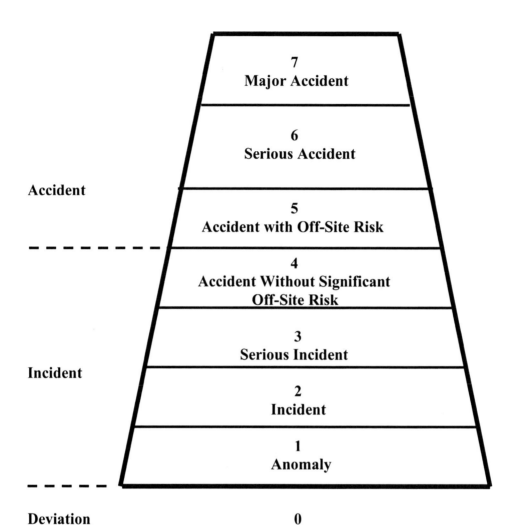

FIGURE 1.1.
International Nuclear Event Scale

WAKE-UP CALLS

On September 29, 1957, a failure of the cooling system for a tank storing tens of thousands of tons of dissolved nuclear waste at Russia's Mayak (Russian for *beacon*; now called Kyshtym) nuclear fuel reprocessing plant resulted in a non-nuclear explosion having a force estimated at about seventy-five tons of TNT, which released some two million curies of radioactivity over 15,000 square miles (American.edu 2009). Subsequently, at least two hundred people died of radiation sickness, ten thousand people were evacuated from their homes, and 470,000 people were exposed to radiation. According to Pollock (1978), people "grew hysterical with fear, with the incidence of unknown 'mysterious' diseases breaking out." Victims were seen with skin "sloughing off their faces, hands and other exposed parts of their bodies." In regard to environmental damage and total body count, Medvedev (1976) points out that "hundreds of square miles were left barren and unusable for decades and maybe centuries. Hundreds of people died, thousands were injured and surrounding areas were evacuated." This accident occurred before the Soviet Union's worst disaster at Chernobyl; it categorized as a level 6 "serious accident" on the 0–7 International Nuclear Events Scale (see figure 1.1).

Only a few days after the Russian Mayak disaster occurred, on October 10, 1957, the graphite core of a British nuclear reactor at Windscale, Cumberland, caught fire, releasing substantial amounts of radioactive contamination into the surrounding area. The event, known as the Windscale fire, was considered the world's worst reactor accident until Three Mile Island in 1979. However, both incidents were dwarfed by the magnitude of the 1986 Chernobyl disaster. The Windscale fire is listed as level 4, accident without significant off-site risk, on the INES scale.

On January 21, 1968, a United States Air Force (USAF) B-52 bomber out of Thule Air Base, Greenland, carrying hydrogen bombs on a routine Cold War alert mission over Baffin Bay caught fire, forcing the crew to abandon the aircraft before it could carry out an emergency landing. Six crew members ejected safely, but one was killed while trying to bail out. The bomber crashed near Thule Air Base in North Star Bay (a.k.a. Bylot Sound), Greenland, causing the nuclear payload to rupture and disperse, which resulted in widespread radioactive contamination. Similar flight operations were discontinued immediately after the incident. Safety procedures were reviewed, and more stable explosives were developed for use in nuclear weapons.

One of the major lessons or tenets of the human experience is that sometimes it takes bad events to bring about positive change. For example, the bad news: consider the events of March 28, 1979, near Middletown, Pennsylvania, when the accident at Three Mile Island Unit 2 (TMI-2)—equipment malfunctions, design-related problems, and worker errors—was the most serious in U.S. commercial nuclear power plant operating history, even though it led to no deaths or injuries to plant workers or members of the nearby community. The good news? TMI-2 brought about sweeping changes involving emergency response planning, reactor operator training, human factors engineering, radiation protection, and many other areas of nuclear power plant operations. It also caused the U.S. Nuclear Regulatory Commission (NRC) to tighten and heighten its regulatory oversight. The bottom line: resultant changes in the nuclear power industry and at the NRC had the effect of enhancing safety (U.S. NRC 2009). This event is listed as a level 5 accident with off-site risk on the INES scale.

At the present time, the only event listed as a level 7 major accident is the Chernobyl accident in 1986 (see the prologue). The nature of events in this accident is best characterized as an external release of a large quantity of the radioactive fuel in the large facility (e.g., the core of the power reactor). This release involved a mixture of short- and long-lived radioactive fission products in quantities radiologically equivalent to more than tens of thousands of terabecquerels (i.e., a unit of radioactivity equal to one disintegration per second) of iodine-131. Such a release would result in the possibility of acute health effect due to iodine-131, which has a relatively short half-life of five days; delayed health effects over a wide area primarily from secondary fission products with longer half-lives of many years, possibly involving more than one country; and long-term environmental consequences.

The incidents described above were all accidents involving civilian or military nuclear power or weapons. To our knowledge, none of these incidents had anything to do with terrorists or terrorism. Again, nuclear power facilities are well protected and very secure from intruders—thankfully, hard nuts to crack. This is not to say they are impenetrable, because nothing we know of is. In the coming chapters we will discuss the basics of protecting industrial facilities, including nuclear facilities, from terrorists and terrorism. At this time, however, it is important to become familiar with terrorism and terrorists.

DID YOU KNOW?

Before the first World Trade Center bombing in 1993 and the 1995 Oklahoma City bombing, emergency incidents were primarily or generally thought to be caused by natural or accidental events.

THE ULTIMATE WAKE-UP CALL

The ultimate wake-up call for Americans came in the form of the events that we now summarize as "9/11" or some characterize as Cantor Fitzgerald 9/11 (because of the 658 employees of this business firm that were killed in the World Trade Center on September 11, 2001). Call it what you will, one fact remains, however: on September 11, 2001, terrorists struck at the heart of America—on American soil in a way that is unforgettable to all of us. Airplanes filled with people and fuel were turned into guided missiles of death and destruction.

The events of 9/11 did not change the way in which we respond to disasters; instead, the tragic events of that day and their aftermath have had a similar impact

Table 1.1. Terror Attack Summaries (within U.S. or against Americans abroad)

Date/Location	Summary of Attacks
Sept. 16, 1920, New York City	Bomb exploded in New York City's Wall Street area, killing forty and injuring hundreds. Perpetrators fled country and were never apprehended.
1951–1956, New York City	Between 1951 and 1956, former Consolidated Edison employee set off series of bombs at New York City landmarks, including Grand Central Station and Radio City Music Hall. No deaths. Bomber was later caught and committed to state mental institution.
Jan. 24, 1975, New York City	Bomb set off in historic Fraunces Tavern killed four and injured more than fifty people. Puerto Rican nationals group (FALN) claimed responsibility, and police tied thirteen other bombings to the group.
Nov. 4, 1979, Tehran, Iran	Iranian radical students seized the U.S. embassy, taking sixty-six hostages. Fourteen were released. The remaining fifty-two were freed after 444 days on the day of President Reagan's inauguration.
1982–1991, Lebanon	Thirty U.S. and other Western hostages kidnapped in Lebanon by Hezbollah. Some were killed, some died in captivity, and some were eventually released. Terry Anderson was held for 2,454 days.
April 18, 1983, Lebanon	U.S. embassy destroyed in suicide car-bomb attack; sixty-three dead, including seventeen Americans. The Islamic Jihad claimed responsibility.
Oct. 23, 1983, Lebanon	Shiite suicide bombers exploded truck near U.S. military barracks at Beirut airport, killing 241 marines. Minutes later a second bomb killed fifty-eight French paratroopers in their barracks in West Beirut.
Dec. 12, 1983, Kuwait City, Kuwait	Shiite truck bombers attacked the U.S. embassy and other targets, killing five and injuring eighty.
Sept. 20, 1984, East Beirut, Lebanon	Truck bomb exploded outside the U.S. embassy annex, killing fourteen, including two U.S. military.
Dec. 3, 1984, Beirut, Lebanon	Kuwait Airways Flight 221, from Kuwait to Pakistan, hijacked and diverted to Tehran. Two Americans killed.
April 12, 1985, Madrid, Spain	Bombing at restaurant frequented by U.S. soldiers, killed eighteen Spaniards and injured eighty-two.
June 14, 1985, Beirut, Lebanon	TWA Flight 847 en route from Athens to Rome hijacked to Beirut by Hezbollah terrorists and held for seventeen days. A U.S. Navy diver executed.
Oct. 7, 1985, Mediterranean Sea	Gunmen attacked Italian cruise ship *Achille Lauro*. One U.S. tourist killed. Hijacking linked to Libya.
Dec. 18, 1985, Rome, Italy, and Vienna, Austria	Airports in Rome and Vienna were bombed, killing twenty people, five of whom were Americans. Bombing linked to Libya.
April 2, 1986, Athens, Greece	A bomb exploded aboard TWA Flight 840 en route from Rome to Athens, killing four Americans and injuring nine.
April 5, 1986, West Berlin, Germany	Libyans bombed a disco frequented by U.S. servicemen, killing two and injuring hundreds.
Dec. 21, 1988, Lockerbie, Scotland	N.Y.-bound Pan-Am Boeing 747 exploded in flight from a terrorist bomb and crashed into Scottish village, killing all 259 aboard and eleven on ground.

Date/Location	Summary of Attacks
Feb. 26, 1993, New York City	World Trade Center bombing in New York City. Six deaths and one thousand injuries. Intended plan of total structural collapse did not occur. Several members of Middle Eastern extremist organizations convicted of roles in the bombing.
Dec. 7, 1993, New York City	Immigrant opened fire on commuters aboard Long Island Railroad commuter train, killing six and wounding nineteen. Gunman convicted of shootings dubbed "the Long Island Railroad Massacre."
April 19, 1995, Oklahoma City	Car bomb exploded outside federal office building, collapsing walls and floors. 168 people were killed, including nineteen children and one person who died in rescue effort. Over 220 buildings sustained damage. Timothy McVeigh and Terry Nichols later convicted in the antigovernment plot to avenge the Branch Davidian standoff in Waco, TX, exactly two years earlier.
Nov. 13, 1995, Riyadh, Saudi Arabia	Car bomb exploded at U.S. military headquarters, killing five U.S. military servicemen.
Oct. 9, 1995, Hyder, Arizona	Twelve-car Amtrak passenger train traveling from New Orleans to Los Angeles carrying 248 people derailed, killing one and seriously injuring twelve. Investigations indicated that tracks had been deliberately tampered with. Incident classified as deliberate act of sabotage. Ongoing investigation in progress.
June 25, 1996, Dhahran, Saudi Arabia	Truck bomb exploded outside Khobar Towers military complex, killing nineteen American servicemen and injuring hundreds of others. Thirteen Saudis and a Lebanese, all alleged members of Islamic militant group Hezbollah, were indicted on charges relating to the attack in June 2001.
July 27, 1996, Atlanta, Georgia	Bomb exploded during concert at Summer Olympics, killing one and injuring 118.
October 1997, San Francisco, CA	An outage at electrical substation in San Francisco affected 126,000 customers. Pacific Gas & Electric said at the time that someone might have deliberately manipulated equipment at the substation to break circuits.
Aug. 7, 1998, Nairobi, Kenya, and Dar es Salaam, Tanzania	Truck bombs exploded almost simultaneously near two U.S. embassies, killing 224 (213 in Kenya and eleven in Tanzania) and injuring about 4,500. Four men connected with al Qaeda, two of whom had received training at al Qaeda camps inside Afghanistan, were convicted of the killings in May 2001 and later sentenced to life in prison. A federal grand jury had indicted twenty-two men in connection with the attacks, including Saudi dissident Osama bin Laden, who remained at large.
May 2000, worldwide impact	Computer hacker in Philippines unleashed "Love Bug" computer virus, overloading corporate and government e-mail systems in many countries. Estimated $10–15 billion in damage. Charges against hacker eventually dropped in Philippines because of inadequate laws against computer hacking.

(continued)

Table 1.1. (*continued*)

Date/Location	Summary of Attacks
July 25, 2000, Detroit, Michigan	Saboteur with knowledge of Detroit lighting system ripped wiring out of one hundred street lights in downtown Detroit, leaving live wires exposed and forcing city officials to shut off power to more than six hundred downtown street lights to ensure public safety. Damages estimated at $26,000. One person was arrested and charge.
September 11, 2001, New York City, Washington, DC, and Pennsylvania	Terrorist cells of Middle Eastern origination hijacked four commercial jet airliners. Two aircraft used against World Trade Center Towers in New York City, third against Pentagon. Fourth aircraft brought down by passenger intervention. Approximately three thousand killed.
September 2001, U.S.	U.S. Postal Service used as delivery vehicle for anthrax spores contained in letters. Numerous news organizations and government agencies affected across U.S. and in some overseas locations with strong U.S. affiliations. Several killed.
October 7, 2001, Alaska	Intoxicated man shoots at the Alaska pipeline, which results in a seven thousand barrel crude oil spill and the pipeline being shut for several days.

Note: Adapted from Infoplease, "Terrorist Attacks," www.infoplease.com/ipa/A0001454.html.

on all aspects of U.S. society by irreversibly changing our beliefs, views, and notions about our modern society. Moreover, what the 9/11 terrorists provided us with was a wake-up call, a general awareness of a type of extremely deadly venom of hate delivered by groups or individuals that is designed to kill massive numbers of people, cause substantial property damage, and affect economic stability. Most importantly, for the public and emergency responders, the events of 9/11 brought to the forefront knowledge and awareness that just about anything is possible. Thus, security of the homeland and our critical infrastructure is vital to all of us. Perhaps nowhere is this point more worrisome in this context than at the nation's 104 operating commercial nuclear power reactors and their adjoining "spent" fuel pools.

Governor Tom Ridge, a U.S. political figure who served as a member of the United States House of Representatives (1983–1995), governor of Pennsylvania (1995–2001), assistant to the president for homeland security (2001–2003), and the first United States secretary of homeland security (2003–2005), got it right when he stated: "You may say Homeland Security is a Y2K problem that doesn't end Jan. 1 of any given year" (Henry 2002). Homeland security is an ongoing problem that must be dealt with 24/7. Simply, there is no magic on-off switch that we can use to turn off the threat of terrorism in the United States or elsewhere.

The threat to our security is not only ongoing but is also universal, including potential and real threats from within—from our own citizens (homegrown terrorism).

Consider the American Timothy McVeigh, for example, who blew up the government building in Oklahoma City in 1995, killing almost two hundred people, including several children. It is interesting to note that McVeigh acted primarily alone. Actually, McVeigh is the exception that proves the rule—most terrorist acts on America are planned by a group beforehand.

Before 9/11, terrorism was not previously seen as a significant threat to the U.S. However, while the U.S. has not experienced continuous 9/11-type terror events in the past, the current climate suggests that the U.S. is at significant risk of further terror activity in the future. Forecasting the future is impossible, but we can look at the past and learn from it. Table 1.1 summarizes past acts of sabotage/terrorism carried out in the U.S. or against U.S. citizens overseas.

Case Study 1.1.   Glowin' in the Dark

My name is W. W. Williams IV . . . just call me Willy or Number Four (I prefer Number Four). You've no doubt heard a recantation similar to the one I'm about to make to y'all right now. Yes, I am one of the Williams boys—the infamous Williams—I call them my blood. For those of you deadheads out there in la-la land who have not heard of my famous bloods, grandfather, father, and now me—all of us in the same fine mess—listen up because I'm goin' to tell you all about it . . . and scare the bejesus out of you!

You see, I come from a long line of the same kind, patriots all, thank you very much! Normally, I would not be having this conversation with you . . . no, sir, I wouldn't. Like my grandfather—the one who blew up that sugar factory—and my father, who had this thing about blowin' up them horrible wind turbines— you know, the ones that kill all them birds and foul up the environment (pun intended), those whirly-gigs that look like crap just a'ruinin' all that pristine wilderness and other landscapes out there. I know you may have heard this before from my deceased grandfather, the one I mentioned earlier, a true patriot, who unfortunately served a very short life sentence; he died a despicable death in a federal prison. So there went another good man, a true patriot, down the proverbial drain, so to speak. And then there is the old man, of course, he is locked up in here, too, but I never see him because they separate us; actually, they separate all of us. I never see anyone but the guards and that crazy shrink and mad doctor now and then.

Now, at the present time, I am disgustingly sober and in total control of my wits and sanity. But like I said, I am not free. No sir. I sit here in my cement coffin, on my cement bed in cement-encased Supermax, Colorado, the ultimate prison for the world's ultimate prisoners. Yes, I am one of the worst; bad to the bone marrow; they do not come any worse, they do not make 'em any worse than me . . . I've never met worse nor will I ever . . . for sure.

(continued)

Sigh! I guess we should get on with my story of what happened and why I am one of the millions of poor souls incarcerated by an American justice system, with those bleeding-heart liberal judges and blood-sucking defense attorneys, that lets all those child molesters and woman beaters go free so they can rape and kill more children and women while honest, patriotic killers like me vegetate in a block of cement like this one.

Anyway, it is what it is—don't you just hate that phrase—anyway, back to the topic and conversation at hand. I was given this chance to state my case, to explain why I put my gang together to build dirty bombs and attack those federal courthouses in a dozen older major cities in the U.S. The idea was to irradiate those areas . . . not to kill people . . . but to scare the bejesus out of them, light 'em in the dark and get our message across. Well, I admit we would have killed a few judges, marshals, lawyers, and a few collaterals here and there, but, hell, that is the price of doing business, you know.

Anyway, let me tell you my story. It ain't pretty in some folks' way of thinking, I suppose. But others' way of thinking has never bothered me too much . . . No, sir. I do it and did it my way. And I would be happy forever sitting on this cement bed speaking to y'all if only my plan had worked.

I joined Greenway Coalition in 1990 during one of those Earth Day celebrations in Boston. I was twelve years old at the time. During my storied career with Greenway, I was involved, along with my father, in some glorious undertakings. For example, I was with my father's team that tree-spiked several old-growth trees on the Olympic Peninsula, Washington State. I personally drove several metal rods into various tree trunks and saved the trees; saving those giants in the old-growth forest was important to me. Unfortunately, a few of those loggers died and others were injured when their chainsaws struck the spikes and the saws kicked back and did other terrible damage, seriously wounding many. The fact is I would kill anyone and everyone to get my point across.

Well, as it turned out, spiking all those trees was great, but looking at all the blood and guts later was a bit overwhelming, so along with my real blood we shifted gears from tree spiking to monkey wrenching or ecotaging.

My blood taught me all I needed to know about monkey wrenching. Our activities in this practice also dealt with saving old-growth forests. However, instead of spiking trees, we ecotaged logging equipment. This turned out to be an ideal practice because I was able to prevent a lot of tree destruction, because the logging equipment all malfunctioned and in some cases massively destroyed itself when they tried to start it. Also, only a few loggers were killed or injured by the monkey-wrenched equipment.

Over the years, my blood's Greenway Coalition grew in numbers and sophistication and I grew right along with it/them. In the late nineties, we shifted gears from forestry to other areas of interest, including committing arson to preserve the pristine wilderness. The Vail Mountain arson attack, for example, is where we torched buildings that were going up. I personally torched one of them horrible, decadent Hollywood homes. All them buildings would have destroyed an outdoor paradise. My blood always told us that there are just some places on earth where man should not intrude. Later we returned to Oregon and torched the main office of a logging company. In 1999 we even went to New Zealand to uproot fields of genetically engineered potatoes. The engineering of foodstuffs should be left to Mother Nature alone and definitely not to man. A few years later we burned down a two-hundred-unit condominium being built in southern California. We put out the word that "if you build it, we will trash it." Then again we shifted gears. We understood that one of the biggest mistakes mankind had made was in developing the automobile. These horrendous machines are nothing but polluters of the environment, with low fuel efficiency and destroyers of things they happen to be run into. So we decided to attack several car dealerships, garages, storehouses in Los Angeles. Most of the autos we attacked were those gas-guzzling, road-hogging, ugly SUVs or Hummers.

A few years ago we decided to change our focus again. We also decided that arson was a bit too damaging . . . we would take out the bad guys for sure, but in some cases we actually burned down the surrounding forest or city blocks, etc. Collateral damage for sure, yes, I admit that, but we decided that we had made our point in that specific area. The fact is we had put a lot of loggers and car dealerships out of business, and the spotted owl sleeps better at night in cleaner air because of our actions.

One of the things that really surprised us, devastated us (if you really must know) was the distorted press and other news media systems out there in la-la land that had the nerve to address all of our heroic actions as acts of vandalism and other forms of destruction. Some even had the audacity to label our actions terrorism or the acts of terrorists. Can you believe that? One thing is certain, the new homeland security secretary, Janet Napolitano, would not be pleased with anyone, including me, being labeled a "terrorist"—she hates that word, bless her for that.

So, here I vegetate in this cement block, locked up like one of those so-called terrorists (no offense intended to the new secretary of homeland security). One media outlet had the audacity to call me crazy, a real nut case, a crank, a weirdo, a jackass, and a cold-blooded baby killer! Can you believe that, man?

Well, let's get back to the story. After my blood was locked up for the wind turbine fiasco . . . and a fiasco it was . . . the fact is, he had hired the wrong gang to do the dirty work; it backfired and he ended up somewhere in this cement tomb. So after my blood was jailed forever, I decided that I needed to take action, and I did.

You see, one thing is for certain, to scare the hell out of Americans all one needs to do is make them think they have been nuked. My plan came to me at one girlfriend's meth labs hidden way back in the hills of western Virginia near a place called the Peaks of Otter. I was up there doing some meth with my sweetheart and

(continued)

her girlfriends and it came over me like a storm, a bolt of lightning illuminating the way for me and my followers. You see, because of my innate ability to plan, my genius instinct to kill and to destroy, I became leader of my own so-called fringe group to be. We had no name or title . . . no, sir. We had no need for any of that. Interference with and destruction of mankind's intrusion on nature was our goal and our mantra.

Back to that stormy night.

Along with the meth, I was tooting up some crack and crank, chased by several shots of my favorite straight bootlegged whiskey. I always drink my whiskey straight—my blood got in trouble for mixing his whiskey and women, so to speak. Anyway, we were lounging around a cabin in the hills that night. During our ethered-up goings on, our conversation centered on who we would kill next—how we could get the biggest bang for the buck. One of the girls worked at a sign factory in the northern Shenandoah Valley. Somehow, someway the factory wench started talking about all the radioactive stuff she worked with in making those exit signs that glow in the dark. She explained that the signs were filled with tritium, a low-grade radioactive gas formed from hydrogen, that along with the materials inside the signs enabled them to glow in the dark; they are a self-powered light source. She went on to explain that the signs and other devices her factory produced were regulated so that when old tritium signs were replaced they had to be returned to her factory or other collection points throughout the country and then shipped to her factory for proper disposal.

During her lengthy explanation of tritium and glow-in-the-dark signs and the required disposal requirements, a thought hit me like a nuclear explosion. I remember thinking: Wow! It is so simple . . . so easy . . . so terrible . . . so wonderful . . . so real . . . so doable . . . so devastating . . . so terrifying . . . so glowing . . . just plain glorious! So we did it.

What we did was keep on talking and listening for a few weeks. We met at the Blue Ridge cabin and did our ether, of course, and I pumped the factory girl for more information. After a few weeks, I had a plan. I decided that I would gather up my team and we would meet at the cabin in one month and discuss my plan.

A month later at the cabin, I had gathered twenty-four of the best female patriots this land has to offer. Many of us had worked together on other ecotage projects, so I had a relatively experienced crew.

That night we gathered around a large campfire and while we did our meth, crack, bootleg whiskey, and other great stuff, I explained the plan.

I told them that we would form into twelve two-person teams. The glow-in-the-dark factory sign worker had told me that she could get her hands on about 3,600 new and used tritium-filled exit signs and could deliver them here, to the cabin site. So that is what she did. I split up the signs among the groups so that each two-person team had three hundred signs. Years earlier, we had begun storing large quantities of high explosives in the woods nearby—we had drawn on this supply many times for various gigs here and there. Anyway, I explained that each team was to take its load of three hundred signs and a sufficient amount of high explosive to make a massive dirty bomb. Again, our intention was not to kill a lot of people but to contaminate twelve major downtown areas.

The first thing we all noticed is that three hundred exit signs is a lot of signs, a big, heavy, bulky load. When we added the plastic explosives to the mix, we knew instantly that two people would not be able to handle such a load; they would not be able to place it where it needed to be placed, either.

So I modified the plan. Instead of bringing more people into the mix to cloud things up and make things more complicated, I decided to work piecemeal. That is, I decided to have each team carry the signs and explosives in increments into the sewers under the courthouses and city halls. Now, we knew this would take at least two hours, but we also knew we had time. Every major target city had extensive underground sewer systems. In my group of twenty-four, I had a couple of gals who had experience in working for sewage districts here and there, so they knew basic underground sewage systems. The plan was to insert the tritium signs and explosives into the sewers in the location to do the most damage to courthouses and city halls or whatever happened to be aboveground. Again, this was to be done by carrying manageable loads of signs and explosives, until all were in place in their intended underground locations.

Obviously, this plan had a lot of moving parts that made it a bit more complicated than I wanted but . . . well, hell, I figured some of the teams would get caught and some would not—it's the price you pay, so to speak. However, it was those that would not get caught that I was counting on. Just a few explosions here and there would get the message across and do horrible damage, which was my goal.

So we set a date and time. I chose a Wednesday, middle of the week—hump day (for sure). I chose a winter night because I wanted dark to come early, and by ten o'clock at night most city centers are empty during the middle of the week.

The teams were dispersed to their target cities with the loads of death and destruction, and all was set. What I did not know (they say what you do not know is what always gets you) is that one of the two-member girl teams, the one up in New England, was of the lazy type. They decided to make their work easier by reducing the tritium signs into one smaller pile that could easily be boxed up and carried into the sewer and placed along with the explosives and detonator at the perfect location. At a remote location outside of Boston, they rented one of those wood shredders, and shredded up all those exit signs, and collected the pulverized materials into a cardboard box, and placed the box into their jeep, and hauled on down the road and into the city. But there was a problem. While driving to town they did not realize that the tritium gas and the powder from the glow-in-the-dark materials contained in the signs was literally covering their entire bodies . . . so they drove down the highway literally glowing in the dark. The first state trooper

(continued)

who happened to see the glowing women speeding down the highway gave chase and pulled them over, handcuffed them and transported them to jail, where they squealed on the rest of us—ratted us out—and the rest is history. Needless to say, none of the dirty bomb gigs actually happened . . . all were arrested and prosecuted by the same system we all hate.

So here I sit, three years later and after one of the longest, most ridiculous trials in criminal history topped off by four losing appeals and waiting on the outcome of another. Like my blood before me, everyone was out to get me. I hadn't killed or injured anyone.

Sitting here now and basking in the total disgrace of failure, only three things bother me: one, I got caught by totally incompetent, dysfunctional deadheads; two, during the trial and all the publicity that followed, they classified me as an American terrorist (Janet Napolitano would not be pleased). "Terrorist?" No way. It is true, I have killed . . . collateral damage . . . but the overall goal of saving the environment is all that mattered . . . the rest is poppycock . . . trivial to the bone. I am not a terrorist. I really don't think the press, law, legal authorities, or the people know what a terrorist really is. Maybe someone ought to write a book on the subject to inform everyone what a terrorist really is. One thing is certain: no, sir, I am not a terrorist. Finally, the third item I am upset about is that I did not stick to my bloods' own mantra. That is, always work alone . . . my main blood always said, "Never, ever mix straight whiskey, you'd turn it sour." Basically, when you mix radiation with meth-heads, with idiots, you light things up . . . you expose yourself and your plans. But you know, I have to admit that it must have been something . . . I would have given anything to see those two meth-heads speeding down the highway . . . just glowin' in the dark.

Before we move on with the rest of the text, it is important to summarize a couple of key points made by the preceding case study. First, the incident just related points to the purpose of this text, namely, to emphasize the importance of critical infrastructure and its protection and being ready for "any" contingency—"any" being the key word. Second, most of the news we listen to or read about at the present time dealing with terrorism, plans about terrorism, or acts of terrorism, is focused on incidents that take place in foreign countries and is perpetrated by foreigners, yet the case study shows a different side of terrorism, domestic terrorism—homegrown terrorism. As described in the case study, the destructive acts were committed by domestic terrorists. This is a touchy area of discussion in the U.S. Americans do not want to visualize or even think about their fellow citizens as being capable of destroying their own country and their own people. However, in reality we feel that terrorist acts committed within the U.S. are most likely to be committed by homegrown individuals than by radical terrorists from other parts of the world. One only need study the history of homegrown terrorism within this country—the acts of the homegrown terrorists Timothy McVeigh, the

Unabomber, and Eric Rudolph—to understand that danger to our country and to its critical infrastructure may be more of an in-house problem, closer to home than a distant, external problem. To retain our country as a free society we must be on guard against all threats, no matter what their origin.

Keep in mind that many define "free society" in different ways. This is important to us in this text because in our view, in order to be truly safe from the terrorist threat we must give up certain freedoms and accept closer scrutiny and vigilance in regard to actions that we normally assume to be no one's business but our own.

At this juncture in the text, it is important to shift gears for a moment and provide a few precise definitions of key terms required for clear understanding of the purview of this text. Keep in mind that the key terms and their definitions presented below may differ slightly (they have been generalized to fit all seventeen of the critical infrastructures) from those used in the U.S. Department of Homeland Security National Infrastructure Protection Plan and the United States Nuclear Regulatory Commission.

Key Terms

Vulnerability is the manifestation of the inherent states of the system (e.g., physical, technical, organizational, social, cultural) that can be exploited by an adversary to adversely affect (cause harm or damage to) that system.

Intent is the desire or motivation of an adversary to attack a target and cause adverse effects.

Capability is the ability and capacity to attack and cause adverse effects.

Threat is the *intent* and *capability* to adversely affect (cause harm or damage to) the system by adversely changing its states.

Risk is the result of a threat with adverse effects to a vulnerable system (Haimes 2004).

WHAT IS TERRORISM?

Since 9/11, we have heard it said by many of our students (and others) that there is controversy about the definition of the politically charged word *terrorism*. The term *terrorism* is a loaded and skewed term. Until the swearing in of the new administration in January 2009, in our opinion, when you get right down to it, the controversy has had more to do with the definition than (until recently) with politics. Terrorism, like pollution, is a judgment call. For example, if two neighbors live next door to an air-polluting facility, one neighbor who has no personal connection with the polluting plant is likely to label the plant's output as pollution. However, the other neighbor, who is an employee of the plant, may consider the plant's pollution as dollar bills— dollars that are his livelihood. On the terrorism front, when someone deliberately spikes a tree to prevent loggers from cutting it down, the tree spiker might feel he is a patriot and not a terrorist. On the other hand, the logger who puts life and limb at risk in taking down the spiked tree has little doubt in his mind about what to call the tree spiker, and it certainly has nothing to do with patriotism. Thus, again, what we are saying here is that along with defining pollution, defining terrorism may be a judgment call, especially in the view of the terrorists.

To make our point on the countless differing views in defining terrorism, consider, for example, that if we were to ask a hundred different individuals to define terrorism, we would likely receive a hundred different definitions. As a case in point, consider the following: if we were to ask a hundred different individuals to describe the actions of W. W. Williams IV (Willy Number Four) in the case study, how would they describe him and his actions? You might be surprised—we certainly were. In 2000 and again in 2003, pre- and post-9/11, after reading Number Four's tritium fiasco, a hundred randomly selected Old Dominion University Environmental Health juniors and seniors ("Generation Y" students ranging in age from twenty to forty years old) were asked to

Table 1.2. Student Responses, Pre- and Post-9/11

Students' Responses Descriptors*	Number of Responses	
	Pre-9/11 (2000)	Post-9/11 (2003)
Question 1: In your opinion, Number Four was		
crazy	30	0
a disgruntled former employee	10	0
insane	14	5
misguided	1	0
a cold-blooded murderer	14	16
a misfit	1	0
deranged	6	4
a lunatic	5	1
a bully	18	5
a terrorist	1	69
not sure	0	0
Totals	100	100
Question 2: In your opinion, Number Four's actions are best described as		
madness	55	1
frustration	9	2
desperation	4	0
dysfunctional thinking	2	0
legitimate concern	1	0
threatening	6	1
terrorism	5	84
workplace violence	0	0
not sure	18	12
Totals	100	100

* Student response descriptors were provided to the students by the instructors.

reply to a nonscientific survey questionnaire. Note that it is common practice to survey our students on pressing environmental and security issues of the day. The two questions and the students' responses to this unscientific survey are listed in table 1.2.

From the Old Dominion University survey it is clear that the students' perceptions of Willy's actions in the case study incident shifted dramatically from pre-9/11 to post-9/11. For example, when asked to select the best pre-9/11 descriptor to describe Number Four, "crazy" and "insane" ranked high; however, after 9/11, the students' perception shifted away from "crazy" and "insane" to "terrorist." Likewise, the students' pre-9/11 responses on Number Four's actions ranked high in "madness"; however, his actions post-9/11 overwhelmingly were described as "terrorism."

It is interesting to note that the year 2000 group reported prior to the September/October 2001 anthrax attacks and prior to 9/11; however, the student group's responses were provided after such events as the World Trade Center attack of 1993 and Timothy McVeigh's 1995 mass murder of the occupants of the federal building in Oklahoma City, Oklahoma. This may help to explain why the year 2000 students were somewhat reluctant to describe Number Four's actions as terrorism and/or to label him as a terrorist.

Terrorism by Any Other Name Is . . .

From the preceding discussion, we might want to buy into the argument that terrorism is relative, a personal judgment. But is it really relative? And if it is relative . . . relative to what? Is it a personal judgment? What is terrorism? Take your choice. Seemingly, there is an endless list of definitions and no universally accepted definition of terrorism. However, notwithstanding the current secretary of homeland security's interpretation of the term *terrorism*, it is terrorism that is the main theme of this text. Let's review a few of these definitions.

Standard Dictionary Definition of Terrorism

After reviewing several dictionaries, we have found this fairly standard definition of terrorism:

> The unlawful use or threatened use of force or violence by a person or an organized group against people or property with the intention of intimidating or coercing societies or governments, often for ideological or political reasons.

America's *National Strategy for Homeland Security* defines terrorism as follows:

> Any premeditated, unlawful act dangerous to human life or public welfare that is intended to intimidate or coerce civilian populations or governments.

The U.S. State Department defines terrorism thus:

> Premeditated, politically motivated violence perpetrated against noncombatant targets by subnational groups or clandestine agents" (USC 2005).

The FBI definition of terrorism is as follows:

> The unlawful use of force or violence against persons or property to intimidate or coerce a Government, the civilian population, or any segment thereof, in furtherance of political or social objectives (FBI 2006).

Note that the FBI divides terrorism into two categories: domestic (homegrown), involving groups operating in and targeting the U.S. without foreign direction, and international, involving groups that operate across international borders and/or have foreign connections.

At this point the obvious question is: Do you now know what terrorism is? That is, can you definitely define it? If you can't define it, you are not alone—not even the U.S. government can definitively define it. Maybe we need to look at other sources—views from the real experts on terrorism.

Osama bin Laden's View on Terrorism

Wherever we look, we find the U.S. as the leader of terrorism and crime in the world. The U.S. does not consider it a terrorist act to throw atomic bombs at nations thousands of miles away [Japan during World War II], when those bombs would hit more than just military targets. Those bombs rather were thrown at entire nations, including women, children, and elderly people . . . (Bergen 2002, 21–22).

Another View

This view is from court testimony on terrorism from Ramzi Ahmed Yousef, who helped organize the first terrorist attack on the World Trade Center:

> You keep talking also about collective punishment and killing innocent people to force governments to change their policies; you call this terrorism when someone would kill innocent people or civilians in order to force the government to change its policies. Well, when you were the first one who invented this. . . .
>
> You were the first one who killed innocent people, and you are the first one who introduced this type of terrorism to the history of mankind when you dropped an atomic bomb which killed tens of thousands of women and children in Japan and when you killed over a hundred thousand people, most of them civilians, in Tokyo with fire bombings.
>
> You killed them by burning them to death. And you killed civilians in Vietnam with chemicals as with the so-called Orange agent. You killed civilians and innocent people,

not soldiers, innocent people every single war you went. You went to wars more than any other country in this century, and then you have the nerve to talk about killing innocent people.

And now you have invented new ways to kill innocent people. You have so-called economic embargo which kills nobody other than children and elderly people, and which other than Iraq you have been placing the economic embargo on Cuba and other countries for over 35 years. . . .

The government in its summations and opening said that I was a terrorist. Yes, I am a terrorist and I am proud of it. And I support terrorism so long as it was against the United States Government and Israel, because you are more than terrorists; you are the one who invented terrorism and using it every day. You are butchers, liars and hypocrites (excerpts from statements in court 1998).

And finally, here is an old cliché on a terrorist:

One man's terrorist is another man's freedom fighter.

Again, from the preceding points of view, it can be seen that defining terrorism or the terrorist is not straightforward and is never easy. Even the standard dictionary definition leaves us with the vagaries and ambiguities of other words typically associated with terrorism, such as in the definitions of *unlawful* and *public welfare* (Sauter and Carafano 2005).

Raphael Perl (2004), in a Congressional Research Service report, points out that one definition widely used in U.S. government circles and incorporated into law defines *international terrorist* as terrorism involving the citizens or property of more than one country.

Terrorism is broadly defined as politically motivated violence perpetrated against noncombatant targets by subnational groups or clandestine agents. For example, kidnapping of U.S. birdwatchers or the bombing of U.S.-owned oil pipelines by guerrillas in Colombia would qualify as international terrorism. In 22 U.S.C. 2656f, a *terrorist group* is defined as a group that practices terrorism or has significant subgroups that practice terrorism. Perl (2004) points out that one of the shortfalls of this traditional definition is its focus on groups and its exclusion of individual ("lone wolf") terrorist activity, which has recently risen in frequency and visibility.

At this point, readers may wonder, "Why should we care—that is, what difference does it make what the definition of terrorist or terrorism is?" Definitions are important because in order to prepare for the terrorism contingency, domestic or international, we must have some feel, as with any other problem, for what it is we are dealing with. We are fighting a war of ideas. We must attempt to understand both sides of the argument, even though the terrorist's side makes no sense to an American or other freedom-loving occupant of the globe.

The *Washington Times* (2009) reports that in an interview with German magazine *Der Spiegel*, the secretary of homeland security said that she's rejected the word *terrorism* and has instead renamed terrorist acts using the euphemism "man-caused disasters," because "it demonstrates that we want to move away from the politics of fear, toward a policy of being prepared for all risks that can occur." The authors of this text suggest that Secretary Napolitano read the remarks we mentioned earlier by Ramzi Ahmed Yousef:

> Yes, I am a terrorist and I am proud of it. And I support terrorism so long as it was against the United States Government and Israel. . . .

Finally, while it is difficult to pinpoint an exact definition of terrorism, we certainly have little difficulty in identifying it when we see it, when we feel it, when we suffer from it. Consider, for example, the earlier account of Number Four's actions in the case study and the various ecotage capers he was involved in. Put yourself in the place of those workers who were harmed or killed by his actions; people who were simply working to support themselves and their families. When the events occurred, none of the victims could have known that an American terrorist had caused the act of terrorism on U.S. soil that killed him/her or their co-workers and/or badly maimed and disfigured others. No, they did not know that. However, if not killed instantly, there is one thing they knew for certain; they knew that crushing feeling of terror as they struggled to breathe, to recover, to survive. By any other name terrorism is best summed up as an absolute feeling of terror—nothing judgmental about that—just terror with a capital *T*.

VOCABULARY OF HATE

In America, there are plenty of hate groups that claim peace and brotherhood, but when their actions are responsible for death and destruction, they are identified for what they really are (Lindsey 2001).

After 9/11, several authors published and the media transmitted seemingly endless accounts of various hate groups operating throughout the globe. Overnight Americans became aware of various theories, philosophies, and terminology very few had ever heard of or thought about. This trend is ongoing—never ending.

Various pundits, so-called experts on the "new" genre of terrorism, have stated that for Americans to understand why foreign terrorists behead innocent people on television or blow up hospitals full of the sick or wounded or schoolhouses full of children, they must get inside the mind of a terrorist—enter the confines of the black box.

The average American might ask: "Get inside the mind of a terrorist? How the hell do you get inside the mind of cold-blooded killers and madmen?"

This is where we make our first mistake, thinking the terrorists act in the manner they do because they are mad, irrational, disturbed, or psychotic. In the case of Timothy McVeigh, we might be able to characterize him and his actions in this manner. Yet McVeigh, as mentioned earlier, is the exception that proves the rule—terrorists' attacks, by real terrorists, are primarily planned beforehand by a group. It is important to remember that McVeigh acted primarily alone.

The terrorists that crashed airplanes into the Twin Towers, Pentagon, and that farm field in Pennsylvania were all of the same mindset; they worked as a group. Likewise, the terrorists that attack Baghdad, Afghanistan, and Pakistan every day work as a group. Terrorists that did all the damage in Bali and Spain and elsewhere acted as a group. Thus, though we would like to classify all the terrorists as we classify Timothy McVeigh, we can't do that. One madman working alone is something we can reasonably assume. However, thinking that hundreds or thousands of like-minded madmen all work in groups is a stretch. The cold-blooded manner in which terrorists go about their business suggests that they are not crazy, insane, or mad, but instead extremely harsh and calculating. If we dismiss them as madmen, we underestimate their intelligence. When we do that, we lose. No, we cannot underestimate the enemy—the terrorists. They are smart, cold-blooded, and calculating. To protect our critical infrastructure, we must be smarter and expect the unexpected—we must be proactive and not just reactive in implementing our countermeasures. Understanding is important. For example, understand that the Koran does not condone murder and suicide; instead, it provides injunctions against suicide and murder. When the Muslim terrorist commits murder or suicide in the name of Islam and the Koran, what he or she is really doing is changing the meaning of the words *murder/suicide* to mean martyrdom. No, getting into the mind of a terrorist is not the solution. Instead, in regard to terrorists, we must ensure they do not get into our minds. Simply, we must be smarter than the enemy.

The Language of Terrorism

Anyone who is going to work at improving the security of America's critical infrastructure must be well versed in the goals and techniques used by the terrorists. Moreover, we cannot implement effective countermeasures unless we know our vulnerabilities. Along with this, we must not only understand what terrorists are capable of doing but also have some feel for their language or vocabulary, which will help us to understand where they are coming from and where they might be headed, so to speak.

As with any other technical presentation, understanding the information presented is difficult unless a common vocabulary is established. Voltaire said it best: "If you wish to converse with me, please define your terms." It is difficult enough to understand terrorists and terrorism (or man-caused disasters); thus, we must be familiar with terms they use and that are used to describe them, their techniques, their materials, and their actions. We have also included some very basic radiation and nuclear power terms and definitions (U.S. NRC 2009); these help set the stage for material that follows.

Definition of Terms

Abu Sayyaf—Meaning "bearer of the Sword"; the smaller of the two Islamist groups whose goal is to establish an Iranian-style Islamic state in Mindanao in the Southern Philippines. In 1991, the group split from the Moro National Liberation Front. With ties to numerous Islamic fundamentalist groups, they finance their operations through kidnapping for ransom, extortion, piracy, and other criminal acts. It is also thought that they receive funding from al Qaeda. It is estimated that there are between two hundred and five hundred Abu Sayyaf terrorists, mostly recruited from high schools and colleges.

acid bomb—A crude bomb made by combining muriatic acid with aluminum strips in a two-liter soda bottle.

aerosol—A fine mist or spray, which contains minute particles.

Afghanistan—At the time of 9/11, Afghanistan was governed by the Taliban, and Osama bin Laden called it home. Amid U.S. air strikes, which began on October 7, 2001, the U.S. sent in more than three hundred million dollars in humanitarian aid. In December 2001, Afghanistan reopened its embassy for the first time in more than twenty years.

aflatoxin—A toxin created by bacteria that grow on stored foods, especially on rice, peanuts, and cottonseeds.

agency—A division of government with a specific function, or a nongovernmental organization (e.g., private contractor, business, etc.) that offers a particular kind of assistance. In the incident command system, agencies are defined as jurisdictional (having statutory responsibility for incident mitigation) or assisting and/or cooperating (providing resources and/or assistance).

airborne—Carried by or through the air.

air marshal—A federal marshal whose purpose is to ride commercial flights dressed in plain clothes and armed to prevent hijackings. Israel's use of air marshals on El Al is credited as the reason Israel has not had a single hijacking in thirty-one years. The U.S. started using air marshals after September 11. However, there are not enough air marshals to go around, so many flights do not have them.

al-Gama'a al-Islamiyya (The Islamic Group, IG)—Islamic terrorist group that emerged spontaneously during the 1970s in Egyptian jails and later in Egyptian universities. After President Sadat released most of the Islamic prisoners from prisons in 1971, groups of militants organized themselves in groups and cells, and al-Gama'a al-Islamiyya was one of them.

al Jazeera—Satellite television station based in Qatar and broadcast throughout the Middle East. Al Jazeera has often been called the "CNN of the Arab world."

alpha radiation—The least penetrating but most damaging type of nuclear radiation. Because of its inability to penetrate the skin, it is not considered dangerous unless ingested.

al Qaeda—Meaning "the base"; an international terrorist group founded in approximately 1989 and dedicated to opposing non-Islamic governments with force and violence. One of the principal goals of al Qaeda was to drive the U.S. armed forces out of the Saudi Arabian peninsula and Somalia by violence. Currently wanted for several terrorist attacks, including those on the U.S. embassy in Kenya and Tanzania as well as the first and second World Trade Center bombings and the attack on the Pentagon.

al Tahwid—A Palestinian group based in London that professes a desire to destroy both Israel and the Jewish people throughout Europe. Eleven al Tahwid were arrested in Germany as they were allegedly about to begin attacking that country.

American Airlines Flight 11—The Boeing 767 carrying eighty-one passengers, nine flight attendants, and two pilots, which was hijacked and crashed into the north tower of the World Trade Center at 8:45 a.m. eastern time on September 11, 2001. Flight 11 was en route to Los Angeles from Boston.

American Airlines Flight 77—The Boeing 757 carrying fifty-eight passengers, four flight attendants, and two pilots, which was hijacked and crashed into the Pentagon at 9:40 a.m. eastern time on September 11, 2001. Flight 77 was en route to Los Angeles from Dulles International Airport in Virginia.

ammonium nitrate-fuel oil (ANFO)—A powerful explosive made by mixing fertilizer and fuel oil. The type of bomb used in the first World Trade Center attack as well as the Oklahoma City bombing.

analyte—The name assigned to a substance or feature that describes it in terms of its molecular composition, taxonomic nomenclature, or other characteristic.

anthrax—An often fatal infectious disease contracted from animals. Anthrax spores have a long survival period, the incubation period is short, and disability is severe, making anthrax a bioweapon of choice by several nations.

antidote—A remedy to counteract the effects of poison.

antigen—A substance that stimulates an immune response by the body's immune system. The body recognizes such substances as foreign and produces antibodies to fight them.

antitoxin—An antibody that neutralizes a biological toxin.

Armed Islamic Group (GIA)—An Algerian Islamic extremist group that aims to overthrow the secular regime in Algeria and replace it with an Islamic state. The GIA began its violent activities in early 1992 after Algiers voided the victory of the largest Islamic party, Islamic Salvation Front (FIS), in the December 1991 elections.

asymmetric threat—The use of crude or low-tech methods to attack a superior or more high-tech enemy.

Axis of Evil—Iran, Iraq, and North Korea, as mentioned by President G. W. Bush during his State of the Union speech in 2002 as nations that were a threat to U.S. security due to harboring terrorism.

Baath Party—The official political party in Iraq until the U.S. debaathified Iraq in May 2003, after a war which lasted a little over a month. Saddam Hussein, the former ruler of the Baath Party, was targeted by American-led coalition forces and fled. Baath Party members have been officially banned from participating in any new government in Iraq.

background radiation—The radiation in the natural environment, including cosmic rays and radiation from the naturally radioactive elements, both outside and inside the bodies of humans and animals. The usually quoted average individual exposure from background radiation is three hundred millirem per year.

Beltway Sniper—For nearly a month in October 2002, the Washington, DC, Maryland, and Virginia area was the hunting grounds for forty-one-year-old John Allen Muhammad and seventeen-year-old Lee Boyd Malvo. Dubbed "the Beltway Sniper" by the media, they shot people at seemingly random places such as schools, restaurants, and gas stations.

bioaccumulative—Substances that concentrate in living organisms rather than being eliminated through natural processes, as in the breather of contaminated air or those who drink contaminated water, live in contaminated areas, or eat contaminated food.

biochemical warfare—Collective term for use of both chemical warfare and biological warfare weapons.

biochemterrorism—Terrorism using as weapons biological or chemical agents.

biological ammunition—Ammunition designed specifically to release a biological agent used as the warhead for biological weapons. Biological ammunition may take many forms, such as a missile warhead or bomb.

biological attacks—The deliberate release of germs or other biological substances that cause illness.

Biosafety Level 1—Suitable for work involving well-characterized biological agents not known to consistently cause disease in healthy adult humans, and of minimal potential hazard to lab personnel and the environment. Work is generally conducted on open bench tops using standard microbiological practices.

Biosafety Level 2—Suitable for work involving biological agents of moderate potential hazard to personnel and the environment. Lab personnel should have specific training in handling pathogenic agents and be directed by competent scientists. Access to the lab should be limited when work is being conducted, extreme precautions should be taken with contaminated sharp items, and certain procedures should be conducted in biological safety cabinets or other physical containment equipment if there is a risk of creating infectious aerosols or splashes.

Biosafety Level 3—Suitable for work done with indigenous or exotic biological agents that may cause serious or potentially lethal disease as a result of exposure by inhalation. Lab personnel must have specific training in handling pathogenic and potentially lethal agents and be supervised by competent scientists who are experienced in working with these agents. All procedures involving the manipulation of infectious material are conducted within biological safety cabinets or other physical containment devices, or by personnel wearing appropriate personal protective clothing and equipment. The lab must have special engineering and design features.

Biosafety Level 4—Suitable for work with the most infectious biological agents. Access to the two Biosafety Level 4 labs in the U.S. is highly restricted.

bioterrorism—The use of biological agents in a terrorist operation. Biological toxins would include anthrax, ricin, botulism, the plague, smallpox, and tularemia.

Bioterrorism Act—The Public Health Security and Bioterrorism Preparedness and Response Act of 2002.

biowarfare—The use of biological agents to cause harm to targeted people either directly, by bringing the people into contact with the agents, or indirectly, by infecting other animals and plants, which would in turn cause harm to the people.

blister agents—Agents that cause pain and incapacitation instead of death and might be used to injure many people at once, thereby overloading medical facilities and causing fear in the population. Mustard gas is the best-known blister agent.

blood agents—Agents based on cyanide compounds. More likely to be used for assassination than for terrorism.

botulism—An illness caused by the botulinum toxin, which is exceedingly lethal and quite simple to produce. It takes just a small amount of the toxin to destroy the central nervous system. Botulism may be contracted by the ingestion of contaminated food or through breaks or cuts in the skin. Food supply contamination or aerosol dissemination of the botulinum toxin are the two ways most likely to be used by terrorists.

Bush Doctrine—The policy that holds responsible nations that harbor or support terrorist organizations and says that such countries are considered hostile to the U.S. From Karen Hughes, White House counselor to the president: "A country that harbors terrorists will either deliver the terrorist or share in their fate.. . . People have to choose sides. They are either with the terrorists, or they're with us."

BWC—Officially known as the "Convention on the Prohibition of Development, Production, and Stockpiling of Bacteriological (Biological) and Toxin Weapons and Destruction." The BWC works toward general and complete disarmament, including the prohibition and elimination of all types of weapons of mass destruction.

Camp X-Ray—The Guantanamo Bay, Cuba, detention camp that houses al Qaeda and Taliban prisoners.

carrier—A person or animal that is potentially a source of infection by carrying an infectious agent without visible symptoms of the disease.

cascading event—The occurrence of one event that causes another event.

causative agent—The pathogen, chemical, or other substance that is the cause of disease or death in an individual.

cell—The smallest unit within a guerrilla or terrorist group. A cell generally consists of two to five people dedicated to a terrorist cause. The formation of cells is born of the concept that an apparent "leaderless resistance" makes it hard for counterterrorists to penetrate.

chain of custody—The tracking and documentation of physical control of evidence.

chemical agent—A toxic substance intended to be used for operations to debilitate, immobilize, or kill military or civilian personnel.

chemical ammunition—A munition, commonly a missile, bomb, rocket, or artillery shell, designed to deliver chemical agents.

chemical attack—The intentional release of toxic liquid, gas, or solid in order to poison the environment or people.

chemical warfare—The use of toxic chemicals as weapons, not including herbicide used to defoliate battlegrounds or riot control agents such as gas or mace.

chemical weapons—Weapons that produce effects on living targets via toxic chemical properties. Examples would be sarin, VX nerve gas, or mustard gas.

chemterrorism—The use of chemical agents in a terrorist operation. Well-known chemical agents include sarin and VX nerve gas.

choking agent—Compounds that injure primarily in the respiratory tract (i.e., nose, throat, and lungs). In extreme cases membranes swell up, lungs become filled with liquid, and death results from lack of oxygen.

Cipro—A Bayer antibiotic that combats inhalation anthrax.

cladding—The thin-walled metal tube that forms the outer jacket of a nuclear fuel rod. It prevents the corrosion of the fuel by the coolant and the release of fission products in the coolants. Aluminum, stainless steel, and zirconium alloys are common cladding materials.

confirmed—In the context of the threat evaluation process, a water contamination incident is confirmed if there is definitive evidence that the water has been contaminated.

containment—The gas-tight shell or other enclosure around a reactor to confine fission products that otherwise might be released to the atmosphere in the event of an accident.

coolant—A substance circulated through a nuclear reactor to remove or transfer heat. The most commonly used coolant in the U.S. is water. Other coolants include graphite, air, carbon dioxide, and helium.

core—The central portion of a nuclear reactor containing the fuel elements and control rods.

counterterrorism—Measures used to prevent, preempt, or retaliate against terrorist attacks.

credible—In the context of the threat evaluation process, a water contamination threat is characterized as credible if information collected during the threat evaluation process corroborates information from the threat warning.

cutaneous—Related to or entering through the skin.

cutaneous anthrax—Anthrax that is contracted via broken skin. The infection spreads through the bloodstream, causing cyanosis, shock, sweating, and finally death.

cyanide agent—Used by Iraq in the Iran war against the Kurds in the 1980s and also by the Nazis in the gas chambers of concentration camps, a cyanide agent is a colorless liquid that is inhaled in its gaseous form, while liquid cyanide and cyanide salts are absorbed by the skin. Symptoms are headache, palpitations, dizziness, and respiratory problems followed later by vomiting, convulsions, respiratory failure, unconsciousness, and eventually death.

cyberterrorism—Attacks on computer networks or systems, generally by hackers working with or for terrorist groups. Some forms of cyberterrorism include denial of service attacks, inserting viruses, or stealing data.

decay heat—The heat produced by the decay of radioactive fission products after the reactor has been shut down.

decontamination—The reduction or removal of contaminating radioactive material from a structure, area, object, or person. Decontamination may be accomplished by (1) treating the surface to remove or decrease the contamination, (2) letting the material stand so that the radioactivity is decreased by natural decay, and (3) covering the contamination to shield the radiation emitted.

dirty bomb—A makeshift nuclear device that is created from radioactive nuclear waste material. While not a nuclear blast, an explosion of a dirty bomb causes localized radioactive contamination as the nuclear waste material is carried into the atmosphere, where it is dispersed by the wind.

Ebola—Ebola hemorrhagic fever (Ebola HF) is a severe, often fatal disease in nonhuman primates such as monkeys, chimpanzees, gorillas, and also in humans. Ebola has appeared sporadically since 1976, when it was first recognized.

eBomb (or e-bomb)—Electromagnetic bomb that produces a brief pulse of energy which affects electronic circuitry. At low levels, the pulse temporarily disables electronics systems, including computers, radios, and transportation systems. High levels completely destroy circuitry, causing mass disruption of infrastructure while sparing life and property.

ecotage—The portmanteau of the "eco-" prefix and "sabotage." It is used to describe illegal acts of vandalism and violence, committed in the name of environmental protection.

ecoterrorism—A neologism for terrorism that includes sabotage intended to hinder activities that are considered damaging to the environment.

emergency feedwater system—Backup feedwater supply used during nuclear plant start-up and shutdown; also known as auxiliary feedwater.

Euroterrorism—Associated with left-wing terrorism of the 1960s, 1970s, and1980s involving the Red Brigade, Red Army Faction, and November 17th Group, among other groups, which targeted American interests in Europe and NATO. Other groups include Orange Volunteers, Red Hand Defenders, Continuity IRA, Loyalist Volunteer Force, Ulster Defense Association, and First of October Anti-Fascist Resistance Group.

fallout—The descent to the earth's surface of particles contaminated with radioactive material from a radioactive cloud. The term can also be applied to the contaminated particulate matter itself.

Fatah—Meaning "conquest by means of jihad"; a political organization created in the 1960s and led by Yasser Arafat. With both a military and intelligence wing, it has carried out terrorist attacks on Israel since 1965. It joined the PLO in 1968. Since 9/11, the Fatah was blamed for attempting to smuggle fifty tons of weapons into Israel.

fatwa—A legal ruling regarding Islamic law.

Fedayeen Saddam—Iraq's paramilitary organization said to be an equivalent to the Nazi's "SS." The militia is loyal to Saddam Hussein and is responsible for using brutality on civilians who are not loyal to the policies of Saddam. They do not dress in uniform.

feedwater—Water supplied to the steam generator that removes heat from the fuel rods by boiling and becoming steam. The steam then becomes the driving force for the turbine generator.

filtrate—In ultrafiltration, the water that passes through the membrane and that contains particles smaller than the molecular weight cutoff of the membrane.

frustration-aggression hypothesis—A hypothesis that every frustration leads to some form of aggression and every aggressive act results from some prior frustration. As defined by Gurr (1968): "The necessary precondition for violent civil conflict is relative deprivation, defined as actors' perception of discrepancy between their value expectations and their environment's apparent value capabilities. This deprivation may be individual or collective."

fuel rod—A long, slender tube that holds fuel (fissional material) for nuclear reactor use. Fuel rods are assembled into bundles called fuel elements or fuel assemblies, which are loaded individually into the reactor core.

fundamentalism—Conservative religious authoritarianism. Fundamentalism is not specific to Islam; it exists in all faiths. Characteristics include literal interpretation of scriptures and a strict adherence to traditional doctrines and practices.

Geneva Protocol 1925—The first treaty to prohibit the use of biological weapons. The 1925 Geneva Protocol for the Prohibition of the Use in War of Asphyxiating, Poisonous or Other Gases and Bacteriological Methods of Warfare.

germ warfare—The use of biological agents to cause harm to targeted people either directly, by bringing the people into contact with the agents, or indirectly, by infecting other animals and plants, which would in turn cause harm to the people.

glanders—An infectious bacterial disease known to cause inflammation in horses, donkeys, mules, goats, dogs, and cats. Human infection has not been seen since 1945, but because so few organisms are required to cause disease, it is considered a potential agent for biological warfare.

grab sample—A single sample collected at a particular time and place that represents the composition of the water, air, or soil only at that time and location.

ground zero—From 1946 until 9/11, ground zero was the point directly above, below, or at which a nuclear explosion occurs or the center or origin of rapid, intense, or violent activity or change. After 9/11, the term, when used with initial capital letters, refers to the ground at the epicenter of the World Trade Center attacks.

guerrilla warfare—The term was invented to describe the tactics Spain used to resist Napoleon, though the tactic itself has been around much longer. Literally, it means "little war." Guerrilla warfare features cells and utilizes no front line. The oldest form of asymmetric warfare, guerrilla warfare is based on sabotage and ambush with the objective of destabilizing the government through lengthy and low-intensity confrontation.

Hamas—A radical Islamic organization that operates primarily in the West Bank and Gaza Strip whose goal is to establish an Islamic Palestinian state in place of Israel. On the one hand, Hamas operates overtly in its capacity as a social services deliverer, but its activists have also conducted many attacks, including suicide bombings, against Israeli civilians and military targets.

hazard—An inherent physical or chemical characteristic that has the potential for causing harm to people, the environment, or property.

hazard assessment—The process of evaluating available information about the site to identify potential hazards that might pose a risk to the site characterization team. The hazard assessment results in assigning one of four levels to risk: lower hazard, radiological hazard, high chemical hazard, or high biological hazard.

hemorrhagic fevers—In general, the term *viral hemorrhagic fever* is used to describe severe multisystem syndrome wherein the overall vascular system is damaged and the body becomes unable to regulate itself. These symptoms are often accompanied by hemorrhage; however, the bleeding itself is not usually life threatening. Some types of hemorrhagic fever viruses can cause relatively mild illnesses.

Hizbollah (Hezbollah)—Meaning "The Party of God." One of many terrorist organizations that seek the destruction of Israel and of the U.S. It has taken credit for numerous bombings against civilians and has declared that civilian targets are warranted. Hizbollah claims it sees no legitimacy for the existence of Israel and that its conflict becomes one of legitimacy that is based on religious ideals.

Homeland Security Office—An agency organized after 9/11, with former Pennsylvania governor Tom Ridge heading it up. The Office of Homeland Security is at the top of approximately forty federal agencies charged with protecting the U.S. against terrorism.

homicide bombings—A term the White House coined to replace the old "suicide bombings."

incident—A confirmed occurrence that requires response actions to prevent or minimize loss of life or damage to property and/or natural resources. A drinking water contamination incident occurs when the presence of a harmful contaminant has been confirmed.

inhalation anthrax—A form of anthrax that is contracted by inhaling anthrax spores. This results in pneumonia, sometimes meningitis, and finally death.

intifada (intifadah)—(alternatively Intifadah, from Arabic "shaking off") The two intifadas are similar in that both were originally characterized by civil disobedience by the Palestinians, which escalated into the use of terror. In 1987, following the killing of several Arabs in the Gaza Strip, the first intifada began and went on until 1993. The second intifada began in September 2000, following Ariel Sharon's visit to the Temple Mount.

Islam—Meaning "submit." The faith practiced by followers of Muhammad. Islam claims more than a billion believers worldwide.

jihad—Meaning "struggle." The definition is a subject of vast debate. There are two definitions generally accepted. The first is a struggle against oppression, whether political or religious. The second is the struggle within oneself, or a spiritual struggle.

kneecapping—A malicious wounding by firearm to damage the knee joint; a common punishment used by Northern Ireland's IRA for those accused of collaborating with the British.

Koran—The holy book of Islam, considered by Muslims to contain the revelations of God to Muhammad. Also called the Qu'ran.

Laboratory Response Network (LRN)—A network of labs developed by the CDC, APHL, and FBI for the express purpose of dealing with bioterrorism threats, including pathogens and some biotoxins.

lassa fever—An acute, often fatal, viral disease characterized by high fever, ulcers of the mucous membranes, headaches, and disturbances of the gastrointestinal system.

LD50—The dose of a substance that kills 50 percent of those infected.

links—The means (road, rail, barge, or pipeline) by which a chemical is transported from one node to another.

mindset—According to *American Heritage Dictionary*: "1. A fixed mental attitude or disposition that predetermines a person's response to and interpretation of situations; 2. an inclination or a habit." *Merriam Webster's Collegiate Dictionary* (10th ed.) defines it as "1. A mental attitude or inclination; 2. a fixed state of mind." The term dates from 1926 but apparently is not included in dictionaries of psychology.

Molotov cocktail—A crude incendiary bomb made of a bottle filled with flammable liquid and fitted with a rag wick.

monkeypox—The Russian bioweapon program worked with this virus, which is in the same family as smallpox. In June 2003, a spate of human monkeypox cases was reported in the U.S. Midwest. This was the first time that monkeypox was seen in North America, and it was the first time that monkeypox was transferred from animal to human. There was some speculation that it was a bioattack.

mullah—A Muslim, usually holding an official post, who is trained in traditional religious doctrine and law.

Muslim (also Moslem)—Followers of the teachings of Muhammad, or Islam.

mustard gas—A blistering agent that causes severe damage to the eyes, internal organs, and respiratory system. Produced for the first time in 1822, mustard gas was not used until World War I. Victims suffered the effects of mustard gas thirty to forty years after exposure.

narcoterrorism—The view of many counterterrorist experts that there exists an alliance between drug traffickers and political terrorists.

National Pharmaceutical Stockpile—A stock of vaccines and antidotes stored at Centers for Disease Control in Atlanta, to be used against biological warfare.

nerve agent—The Nazis used the first nerve agents: insecticides developed into chemical weapons. Some of the better-known nerve agents include VX, sarin, soman, and tabun. These agents are used because only a small quantity is necessary to inflict substantial damage. Nerve agents can be inhaled or can absorb through intact skin.

nodes—A facility at which a chemical is produced, stored, or consumed.

nuclear blast—An explosion of any nuclear material that is accompanied by a pressure wave, intense light and heat, and widespread radioactive fallout, which can contaminate the air, water, and ground surface for miles around.

nuclear reactor—A device in which nuclear fission may be sustained and controlled in a self-supporting nuclear reaction. There are several varieties, but all incorporate certain features, such as fissionable material or fuel, a moderating material (to control the reaction), a reflector to conserve escaping neutrons, provisions for removal of heat, measuring and controlling instruments, and protective devices.

opportunity contaminant—A contaminant that might be readily available in a particular area, even though it may not be highly toxic or infectious or easily dispersed and stable in treated drinking water.

Osama bin Laden (also spelled "Usama")—A native of Saudi Arabia, he was born the seventeenth of twenty-four sons of Saudi Arabian builder Mohammed bin Oud bin Laden, a Yemeni immigrant. Early in his career, he helped the mujahideen fight the Soviet Union by recruiting Arabs and building facilities. He hates the U.S. and apparently this is because he views the U.S. as having desecrated holy ground in Saudi Arabia with its presence during the first Gulf War. Expelled from Saudi Arabia in 1991 and from Sudan in 1996, he operated terrorist training camps in Afghanistan. His global network al Qaeda is credited with the attacks on the U.S. on September 11, 2001, the attack on the USS *Cole* in 2000, and a number of other terrorist attacks.

pathogen—Any agent that can cause disease.

pathways—The sequences of nodes and links by which a chemical is produced, transported, and transformed from its initial source to its ultimate consumer.

plague—The pneumonic plague, which is more likely to be used in connection with terrorism, is naturally carried by rodents and fleas but can be aerosolized and sprayed from crop dusters. A 1970 World Health Organization assessment asserted that in a worst case scenario, a dissemination of fifty kilograms in an aerosol over a city of five million could result in 150,000 cases of pneumonic plague, 80,000–100,000 of which would require hospitalization and 36,000 of which would be expected to die.

political terrorism—Terrorist acts directed at governments and their agents and motivated by political goals (e.g., national liberation).

possible—In the context of the threat evaluation process, a water contamination threat is characterized as possible if the circumstances of the threat warning appear to have provided an opportunity for contamination.

potassium iodide—An FDA-approved nonprescription drug for use as a blocking agent to prevent the thyroid gland from absorbing radioactive iodine.

pressure vessel—A strong-walled container housing the core of most types of power reactors.

pressurizer—A tank or vessel that controls the pressure in a certain type of nuclear reactor.

presumptive results—Results of chemical and/or biological field testing that need to be confirmed by further lab analysis. Typically used in reference to the analysis of pathogens.

primary system—The cooling system used to remove energy from the reactor core and transfer that energy either directly or indirectly to the steam turbine.

psychopath—A mentally ill or unstable person, especially one having a psychopathic personality (*see* psychopathy), according to *Webster's*.

psychopathology—The study of psychological and behavioral dysfunction occurring in mental disorder or in social disorganization, according to *Webster's*.

psychopathy—A mental disorder, especially an extreme mental disorder marked usually by egocentric and antisocial activity, according to *Webster's*.

psychotic—Of, relating to, or affected with psychosis, which is a fundamental mental derangement (as schizophrenia) characterized by defective or lost contact with reality, according to *Webster's*.

radiation—Particles (alpha, beta, neutrons) or photons (gamma) emitted from the nucleus of an unstable atom as a result of radioactive decay.

rapid field testing—Analysis of water during site characterization uses rapid field water testing technology in an attempt to tentatively identify contaminants or unusual water quality.

red teaming—As used in this text, a group exercise to imagine all possible terrorist attack scenarios against the nuclear infrastructure and their consequences.

retentate—In ultrafiltration, the retentate is the solution that contains the particles that do not pass through the membrane filter. The retentate is also called the concentrate.

ricin—A stable toxin easily made from the mash that remains after processed castor beans. At one time, it was used as an oral laxative, castor oil; ricin causes diarrhea, nausea, vomiting, abdominal cramps, internal bleeding, liver and kidney failure, and circulatory failure. There is no antidote.

salmonella—An infection caused by a gram-negative bacillus, a germ of the *Salmonella* genus. Infection with this bacteria may involve only the intestinal tract or may be spread from the intestines to the bloodstream and then to other sites in the body. Symptoms of salmonella enteritis include diarrhea, nausea, abdominal pain, and fever. Dehydration resulting from the diarrhea can cause death, and the disease could cause meningitis or septicemia. The incubation period is between eight and forty-eight hours, while the acute period of the illness can hang on for one to two weeks.

sarin—A colorless, odorless gas. With a lethal dose of .5 milligrams (a pinprick-sized droplet), it is twenty-six times more deadly than cyanide gas. Because the vapor is heavier than air, it hovers close to the ground. Sarin degrades quickly in humid weather, but sarin's life expectancy increases as the temperature gets higher, regardless of how humid it is.

secondary system—The steam generator tubes, steam turbine, condenser and associated pipes, pumps, and heaters used to convert the heat energy of the reactor coolant system into mechanical energy for electrical generation.

sentinel laboratory—A Laboratory Response Network (LRN) lab that reports unusual results that might indicate a possible outbreak and refers specimens that may contain select biological agents in reference labs within the LRN.

site characterization—The process of collecting information from an investigation site in order to support the evaluation of a drinking water contamination threat. Site characterization activities include the site investigation, field safety screening, rapid field testing of the water, and sample collection.

sleeper cell—A small cell that keeps itself undetected until such time as it can "awaken" and cause havoc.

smallpox—The first biological weapon, used during the eighteenth century, smallpox killed three hundred million people in the nineteenth century. There is no specific treatment for smallpox disease, and the only prevention is vaccination. This currently poses a problem, since the vaccine was discontinued in 1970 and the WHO declared smallpox eradicated. Incubation is seven to seventeen days, during which the carrier is not contagious. Thirty percent of people exposed are infected, and it has a 30 percent mortality rate.

sociopath—Basically synonymous with psychopath. Sociopathic symptoms in the adult sociopath include an inability to tolerate delay or frustration, a lack of guilt feelings, a relative lack of anxiety, a lack of compassion for others, a hypersensitivity to personal ills, and a lack of responsibility. Many authors prefer the term *sociopath* because this type of person had defective socialization and a deficient childhood.

sociopathic—Of, relating to, or characterized by asocial or antisocial behavior or a psychopathic personality, according to *Webster's*.

spore—An asexual, usually single-celled reproductive body of plants such as fungi, mosses, or ferns; a microorganism, as a bacterium, in a resting or dormant state.

steam generator—The heat exchanger used in some reactor designs to transfer heat from the primary (reactor coolant) system to the secondary (steam) system. This design permits heat exchange with little or no contamination of the secondary system equipment.

terrorist group—A group that practices or has significant elements that are involved in terrorism.

threat—An indication that a harmful incident, such as contamination of the drinking water supply, may have occurred. The threat may be direct, such as a verbal or written threat, or circumstantial, such as a security breach or unusual water quality.

toxin—A poisonous substance produced by living organisms capable of causing disease when introduced into the body tissues.

transponder—A device on an airliner that sends out a signal allowing air traffic controllers to track the aircraft. Transponders were disabled in some of the planes hijacked 9/11.

Transportation Security Administration (TSA)—A new agency created by the Patriot Act of 2001 for the purpose of overseeing technology and security in American airports.

tularemia—An infectious disease caused by a hardy bacterium, *Francisella tularensis*, found in animals, particularly rabbits, hares, and rodents. Symptoms depend upon how the person was exposed to tularemia but can include difficulty breathing, chest pain, bloody sputum, swollen and painful lymph glands, ulcers on the mouth or skin, swollen and painful eyes, and sore throat. Symptoms usually appear from three to five days after exposure but sometimes take up to two weeks. Tularemia is not spread from person to person, so people who have it need not be isolated.

turbine—A rotary engine made with a series of curved vanes on a rotating shaft. Usually turned by water or steam. Turbines are considered to be the most economical means to turn large electrical generators.

ultrafiltration—A filtration process for water that uses membranes to preferentially separate very small particles that are larger than the membrane's molecular weight cutoff, typically greater than 10,000 daltons. (A dalton is a unit of mass, defined as one-twelfth the mass of a carbon-12 nucleus. It's also called the atomic mass unit, abbreviated as either "amu" or "u").

vector—An organism that carries germs from one host to another.

vesicle—A blister filled with fluid.

weapons of mass destruction (WMD)—According to the National Defense Authorization Act, any weapon or device that is intended, or has the capability, to cause death or serious bodily injury to a significant number of people through the release, dissemination, or impact of (a) toxic or poisonous chemicals or their precursors, (b) a disease organism, or (c) radiation or radioactivity.

xenophobia—Irrational fear of strangers or those who are different from oneself.

zyklon b—A form of hydrogen cyanide. Symptoms of inhalation include increased respiratory rate, restlessness, headache, and giddiness followed later by convulsions, vomiting, respiratory failure, and unconsciousness. Used in the Nazi gas chambers in World War II.

REFERENCES AND RECOMMENDED READING

American.edu 2009. *List of military nuclear accidents*. www.american.edu/projects/mandala/TED/ural.htm (accessed October 5, 2009).

Bergen, P. L. 2002. *Holy War, Inc: Inside the secret world of Osama bin Ladin*. New York: Touchstone Press, 21–22.

CRS. 2006. *Chemical facility security. CRS report for Congress*. Washington, DC: Congressional Research Service—Library of Congress.

Diaz, N. 2004. Quoted in *Nuclear plant security after Sept. 11: Quotes from experts*. NEI, www.nei.org (accessed October 5, 2009).

Excerpts from statements in court. 1998. *New York Times*, January 9, B4.

FBI. 2006. *Terrorism 2002–2005*. Washington, DC: Federal Bureau of Investigation. www.fbi
.gov/publications/terror/terrorism 2000_2005.htm (accessed January 5, 2010).

Gurr, T. R. 1968. Psychological factors in civil violence. *World Politics* 20: 245–78.

Haimes, Y. Y. 2004. *Risk modeling, assessment, and management*. 2nd ed. New York: John
Wiley & Sons, 699.

Henry, K. 2002. New face of security. *Government Security*, April 1, 30–37.

Lindsey, H. 2001. *Vocabulary of hate*. www.wordnetdaily.com (accessed April 18, 2008).

Medvedev, Z. 1976. Mayak. *Australian* (September 12).

Napolitano tells it like it isn't. *Washington Times*. www.washingtontimes.com/news/2009/
mar/29/tell-it-like-it-is-man-caused-disasters-is-napolit/ (accessed March 29, 2009).

NEI. 2009. *Safety & security*. NEI, http://www.nei.org/keyissues/safetyandsecurity (accessed
October 1, 2009).

NSHS. 2006. *National strategy for homeland security*. www.whitehouse/homeland (accessed
May 13, 2006).

Old Dominion University. 2000; 2002. *Violence in the workplace: Security concerns*. From a
series of lectures presented to environmental health students. Norfolk, VA.

OSHA. 2007. *Combustible dust national emphasis program: CPL 03-00-006*. Washington,
DC: U.S. Department of Labor. www.osha.gov/pls/oshaweb/owadisp.show_document?p_
table=DIRECTIVES&p_id=3729 (accessed April 14, 2008).

Perl, R. 2004. *Terrorism and national security: Issues and trends*. CRS Issue Brief IB10119.
Washington, DC.

Pollock, R. 1978. Soviets experience nuclear accident, *Critical Mass Journal* 3, no. 10: 9.

Sauter, M. A., and Carafano, J. J. 2005. *Homeland security: A complete guide to understanding,
preventing, and surviving terrorism*. New York: McGraw-Hill.

Spellman, F.R., (1997). *A Guide to Compliance for Process Safety Management/Risk
Management Planning (PSM/RMP)*. Lancaster, PA: Technomic Publishing Company.

USC (United States Congress). 2005. Annual country reports on terrorism. 22 USC, chapter
38, section 2656f.

U.S. NRC. 2009. *Backgrounder on the Three Mile Island accident*. http://www.nrc.gov/reading
-rm/doc-collections/fact-sheets/3mile-isle.html (accessed October 6, 2009).

2

Critical Infrastructure

While it is not so easy to definitively define terrorism and/or the terrorist, we have less difficulty identifying the likely targets of terrorists. In America, we call these likely targets our critical infrastructure.

If anything, the NRC could be faulted for overkill, as nuclear power plants have always been extremely secure. Worst-case scenarios of terrorist attacks on plants or nuclear waste under transport indicate a very low likelihood of collateral injury.

—*America At Risk: A Homeland Security Report Card, Progressive Policy Institute, Democratic Leadership Council, July 2003*

WHAT IS CRITICAL INFRASTRUCTURE?

For the United States of America, 9/11 was a slap in the face, a punch in the gut (the ultimate sucker punch), and a most serious wake-up call. The 9/11 wake-up call generated several reactions on our part—protecting ourselves from further attack became (and, we hope, still is) priority number one. In light of this important need (i.e., the survival of our way of life), the Department of Homeland Security was created. According to Governor Tom Ridge, "You may say Homeland Security is a Y2K problem that doesn't end Jan. 1 of any given year" (Henry 2002). And, according to Barack Obama (2007), the Department of Homeland Security does "the work that ensures no other family members have to lose a loved one to a terrorist who turns a plane into a missile, a terrorist who straps a bomb around her waist and climbs aboard a bus, a terrorist who figures out how to set off a dirty bomb in one of our cities."

Among other things, the new emphasis on homeland security pointed to the need to protect and enhance the security of the nation's critical infrastructure. Critical infrastructure can be defined or listed in many ways. Generally, governments use the term to describe material assets, systems, and services that are essential for the functioning of an economy and society and maintaining public confidence. Destruction or compromise of any of these systems or services would have a debilitating impact on the area either directly, through interdependencies, or from cascading effects. For the purpose of this text, critical infrastructure is defined as those assets of physical, key resources and computer/service-based systems that are essential to the minimum operations of an economy and government. The authors define critical infrastructures as the following:

- agriculture
- banking and finance
- chemical and hazardous materials
- defense industrial base
- emergency services
- energy
- organizations
- postal and shipping
- public health
- strategies and assessments
- telecommunications
- transportation
- water and water treatment systems
- dams
- information technology
- national monuments and icons
- commercial assets
- nuclear power plants

DID YOU KNOW?

Over 85 percent of the critical infrastructure within the U.S. is owned and operated by the private sector.

Although we did not list cyberspace and all ancillaries (excluding the listing of information technology) involved in or with digital operations (e-technology), in this current era we can state without equivocation, doubt, ambiguity, or vagueness that the digital connection is the glue that holds all critical infrastructure together. This is the case because all separate infrastructures are interconnected in one way or another. It would be hard to imagine that any of the above listed infrastructure sectors could operate today without e-technology.

For example, consider e-agriculture. The Food and Agriculture Organization of the United Nations (FAO 2005) defines e-agriculture as "an emerging field within agricultural informatics, agricultural development and business," referring to "agriculture services and information delivered or enhanced through the Internet and related technologies." More specifically, it involves the "conceptualization, design, development, evaluation and application of new (innovative) ways to use existing or emerging information and communication technologies."

DID YOU KNOW?

More advanced applications of e-agriculture in farming exist in the use of sophisticated information and communication technologies such as satellite systems, global positioning systems (GPS), advanced computers, and electronic systems to improve the quantity and quality of production (FAO 2005).

Even though we did it in the past, today how would we go about withdrawing money from the bank without e-banking technology? Today we can conduct our banking at any time we wish, from any location in the world. Modern life without debit cards and ATMs would be a rude awakening for many of us.

We could go down the list of critical infrastructures and easily point out where e-technology and the specific industry interface. However, since this is a discussion focusing on the energy industry, we will keep the focus on e-technology as it relates to or interfaces with the energy industry. If not familiar with the energy industry—an oil refinery, hydroelectric dam, or wind farm, for example—it might surprise you to know that the entire modern oil refining process and electrical distribution grid is operated by e-technology from computer operation stations (manned by people) with various digital proximity switches and other devices strategically positioned throughout the process to operate valves and switches, monitor critical parameters, and provide automatic emergency shutdown procedures.

NUCLEAR POWER INFRASTRUCTURE*

In the first four volumes of our critical infrastructure series, *Water Infrastructure Protection and Homeland Security, Food Infrastructure Protection and Homeland Security, The Chemical Industry and Homeland Security,* and *The Energy Industry and Homeland Security,* the message/focus was on water/wastewater, agriculture, and chemical manufacturing, processing, and storage, and conventional energy producers oil, gas, coal, and electricity, and energy power utilization in the United States. In this fifth book of the series, *Nuclear Infrastructure Protection and Homeland Security,* even though the target is different—pointing out and discussing the threat to our nuclear industry (nuclear reactors, materials, and waste)—we use the same proven format. In addition, this text describes the study, design, and implementation of precautionary measures aimed to reduce the risk to our nuclear infrastructure from both homegrown and/or foreign terrorism.

The U.S. Department of Homeland Security (2009) points out that energy infrastructure fuels the economy of the twenty-first century. Without a stable energy supply, health and welfare are threatened and the U.S. economy can't function. Energy infrastructure is divided into six interrelated segments: electricity, petroleum, coal, natural gas, emerging renewable energy (biomass, hydro, wind, solar, tides, etc.), and nuclear power. The first four segments were discussed in detail in the previously mentioned member of the critical infrastructure series *The Energy Industry and Homeland Security.*

National Infrastructure Protection Plan

Homeland Security Presidential Directive 7 (HSPD-7) identified critical infrastructure and key resource (CIKR) sectors and designated federal government sector-specific agencies (SSAs) for each of the sectors. Each sector is responsible for developing and implementing a sector-specific plan (SSP) and providing sector-level performance feedback to the Department of Homeland Security (DHS) to enable gap assessments of national cross-sector CIKR protection programs. Collaboration with private-sector partners and encouraging the development of appropriate information sharing and analysis within each sector is critical.

Nuclear Power Sector

Approximately 20 percent of the nation's electrical power is derived from the 104 commercial nuclear reactors licensed to operate in the United States. The nuclear reactors, materials, and (nuclear) waste sector includes nuclear power plants; nonpower

---

* Much of the information in this section is from Department of Homeland Security's *National Infrastructure Protection Plan: Energy Sector* and *National Infrastructure Protection Plan: Nuclear Reactors, Materials, and Waste Sector.* Retrieved 2009 from www.dhs.gov/nipp.

nuclear reactors used for research, testing, and training; nuclear materials used in medical, industrial, and academic settings; nuclear fuel fabrication facilities; decommissioning reactors; and the transportation, storage, and disposal of nuclear material and waste.

The nuclear sector has identified interdependencies with other critical infrastructure sectors, including

- **energy**—as a supplier to the nation's electrical grid
- **transportation systems**—through the movement of radioactive material
- **chemical**—related to hazardous chemicals at fuel cycle facilities
- **public health and health care**—through nuclear medicine, radiopharmaceuticals, and sterilization of surgical supplies
- **government facilities**—through federal and state facilities that use radioactive material for various purposes

HSPD-7 assigned responsibility for the protection of the nuclear sector to the Department of Homeland Security (DHS). Within DHS, the Sector-Specific Agency Executive Management Office (SSA EMO) and, specifically, the Nuclear Sector-Specific Agency will maintain responsibility for CIKR protection of the sector, in close cooperation with the Nuclear Regulatory Commission (NRC) and the Department of Energy (DOE).

DHS established government and private-sector coordinating councils in 2004 to plan and coordinate CIKR protection efforts for the sector. These councils provide a structure through which representative groups from all levels of government and the private sector can collaborate and share approaches to CIKR protection. The Government Coordinating Council consists of representatives from DHS, NRC, DOE, the Environmental Protection Agency, the Federal Bureau of Investigation, and the Department of State, as well as representatives from several state nuclear regulatory agencies and organizations. The Nuclear Government Coordinating Council (GCC) coordinates civilian nuclear security strategies, activities, policies, and communications across and between the government and the nuclear sector. The Nuclear Sector Coordinating Council consists of representatives from the nuclear industry who collaborate to share information and concerns regarding CIKR protection.

Nuclear Power/Materials Security Issues

One of the primary goals of terrorists is to instill fear in as many people as possible. It is important to remember that an offshoot of terrorism is terror. While the loss of electricity generated by a single nuclear power plant may have only a minor impact on the nation's overall electrical capacity, a terrorist attack on a nuclear power plant or research reactor would be a significant security event, especially if a successful ter-rorist strike resulted in the release of radioactive material. For this reason, nuclear sector facilities are among the best defended and most physically hardened of the na-tion's CIKR, designed to withstand such extreme events as hurricanes, tornadoes and tornado-generated missiles, and earthquakes.

Another area of concern is nuclear materials. These source materials are used in a variety of medical and industrial settings and licensed by the NRC. Attacks may involve seizure of a nuclear vessel or facility or theft of radioactive materials. If stolen or otherwise misappropriated, these materials could be used in a radiological dispersal device (RDD, or "dirty bomb") or a radiological exposure device (RED). GCC mem-bers are therefore working through a variety of programs and initiatives to ensure the radioactive sources are used only as intended.

Nuclear Sector Security and Preparedness Program

Priority protective programs in the nuclear sector are operated by DHS, DOE, NRC, and other sector partners. Some current nuclear sector security and prepared-ness programs include the following:

- **Comprehensive Review (CR) Program**—This program is a cooperative federal, state, local, and private-sector analysis of CIKR facilities to determine the security and response capabilities of the facilities and their surrounding community. The nuclear sector recently completed comprehensive reviews at all of the nation's nuclear power plants. DHS coordinated this interagency effort and is working with its partners to address potential enhancements identified during the CR process.
- **Buffer Zone Plans (BZPs)**—These plans identify and recommend security measures and local law enforcement coordination for the area surrounding a facility (the "buf-fer zone"), making it more difficult for a potential attacker to conduct surveillance or to plan or launch an attack. DHS has set aside grant funds to address potential security enhancements identified through the CR assessment process. Grants issued under this program will be strictly risk based and carefully targeted as creating or reinforcing specific capabilities in the communities surrounding critical infrastruc-ture and key resources.
- **Radiological Emergency Preparedness (REP) Program**—This is a FEMA program that provides oversight of radiological emergency planning and preparedness activi-ties. REP leads offsite emergency planning and reviews and evaluates radiological emergency response plans and procedures developed by state and local govern-ments. This program serves to enhance planning, preparedness, and response for all types of peacetime radiological emergencies with federal, state, and local govern-ments and the private sector.

- **Atomic Energy Act**—This 1954 act (a substantial amendment to the 1946 statute) ensures proper management, safety, and security of the nation's atomic energy by providing a program for government control of the possession, use, and production of atomic energy and specific nuclear materials, whether owned by the government or the private sector. The act assigns control of special nuclear materials, source materials, and by-product materials to the NRC. It requires licensing of civilian uses of nuclear materials and facilities and empowers the NRC to establish and enforce standards to govern the use of such materials. While nuclear sector facilities are largely run by private companies, the NRC is charged with ensuring that the facilities are operated according to NRC rules and regulations.

THE BOTTOM LINE

Again, it is important to point out that the nuclear sector provides products and materials that are essential to the U.S. economy and to the standard of living we currently enjoy. The nuclear sector is well aware of its vulnerabilities and is leading a significant voluntary and regulated effort to increase its planning and preparedness. Cooperation through industry groups has resulted in substantial information sharing of effective and best practices across the sector. Many sector owners and operators have extensive experience with infrastructure protection and have more recently focused their attention on cyber security. In addition to the economic consequences of a successful homegrown or foreign terrorist attack against nuclear sector facilities, there is also the potential of a threat to public health and safety and the environment. We hope that this book and the others in the critical infrastructure series will aid in the prevention and mitigation of deliberate attacks, whether foreign or domestic.

REFERENCES AND RECOMMENDED READING

Breeze, R. 2004. Agroterrorism: Betting far more than the farm. *Biosecurity and Bioterrorism: Biodefense Strategy, Practice and Science* 2(4), 1–14.

Carus, S. 2002. *Bioterrorism and biocrimes: The illicit use of biological agents since 1900.* Washington, DC: Center of Counterproliferation Research, National Defense University.

CBO. 2004. *Homeland security and the private sector.* Washington, DC: Congressional Budget Office.

Chalk, P. 2004. *Hitting America's soft underbelly: The threat of deliberate biological attacks against the U.S. agriculture and food industry.* Santa Monica, CA: RAND.

FAO. 2005. *Bridging the rural digital divide.* New York: United Nations. www.fao.org/rdd (accessed April 19, 2008).

Federal Register. 2003. Notice of proposed rulemaking. *Federal Register* 68(90).

Henry, K. 2002. New face of security. *Government Security,* April 1, 30–37.

Horn, F. P. 1999. Statement made before the United States Senate Emerging Threats and Capabilities Subcommittee of the Armed Services Committee. armed_services.senate.gov/statemnt/1999/991027fh.pdf (accessed June 27, 2007).

Lane, J. 2002. Sworn testimony, Congressional Field Hearing, House Committee on Government Reform, Abilene, KS.

Parker, H. S. 2002. *Agricultural bioterrorism: A federal strategy to meet the threat.* McNair Paper 65, Nation Defense University. http://www.ndu.edu/inss/McNair/mcnair65/McN_65.

Obama, B. 2007. *Homeland security.* http://www.whitehouse.gov/agenda/homeland_security/.

U.S. Department of Homeland Security. 2009. *National infrastructure protection plan: Energy sector.* www.dhs.gov/nipp (accessed April 24, 2009).

U.S. Nuclear Power Sector

Question: Why nuclear power?

Answer: Nuclear power produces no CO2 emissions, no nitrous oxide emissions, no sulfuric emissions; without emissions, no headlines.

Nuclear power plants have become of great concern to everyone. But they are probably our best defended targets. There is more security around nuclear power plants than anything else we've got. When confronted by an aviation attack, there is some concern. But one of the things that we have clearly found in this exercise is that this is an industry that has taken security pretty seriously for quite a long time, and its infrastructure, especially against these kinds of terrorist threats, is extremely good.

—*John Hamre, President and CEO, Center for Strategic and International Studies,*
from "Silent Vector," a national security simulation, October 2002

INTRODUCTION*

In the fourth volume of this series, it was pointed out that the United States energy sector is vital to the U.S. economy. It is a high-tech, research and development (R&D)–oriented industry that, as mentioned, includes assets related to three key conventional energy resources: electric power, petroleum, and natural gas. Each of these resources requires a unique set of supporting activities and assets, as shown in table 3.1. Petroleum and natural gas share similarities in methods of extraction, fuel cycles, and transport, but the facilities and commodities are separately regulated and have multiple stakeholders and trade associations.

---

* Much of the information in this section is from Department of Homeland Security (2007), *Energy: Critical infrastructure and key resources sector-specific plan as input to the national infrastructure protection plan* (Washington, DC); Energy Information Administration (2009a), Nuclear power and the environment; Nuclear explained; Introduction to nuclear power, http://www.eia.doe.gov/cneaf/nuclear/page/nuclearenvissues.html (accessed October 19, 2009).

Table 3.1. Segments of the Energy Sector

Electricity	Petroleum	Natural Gas
Generation	Crude oil	Production
Transmission	Petroleum processing facilities	Transport
Distribution		Distribution
Control systems		Storage
Electricity markets		Liquefied natural gas
		Control systems
		Gas markets

Energy assets and critical infrastructure components are owned by private, federal, state, and local entities, as well as by some types of energy consumers, such as large industries and financial institutions (often for backup power purposes). Types of major asset ownership are shown in table 3.2.

Even though Americans have used conventional energy supplies more wisely and efficiently since the oil embargo of 1973, our population has continued to grow to more than 310 million, and it is still growing. Moreover, our dependence on foreign (not always reliable) sources of oil has continued to grow, increasing daily. While we have an estimated supply of home-mined coal to last more than two centuries, coal is a dirty source of energy that contributes to environmental pollution. It is true that coal-powered plants can be made to burn coal more cleanly, but it is expensive to install and not always efficient in operation. At present, U.S. natural gas supplies are abundant and will be a reliable source of energy for the rest of the twenty-first century.

Table 3.2. Major Asset Ownership

Ownership Entities	Assets
Federal government	Hydroelectric dams, nuclear and fossil fuel power generation stations, and high-voltage transmission
State and local government	All municipal utilities
Regulated utilities	Most of the electric and natural gas infrastructure in the U.S., including major interstate pipeline companies, hydroelectric facilities, storage facility operators, and LNG terminal owners
Unregulated energy companies	Energy infrastructure assets, such as merchant generation companies owning power plants that participate in wholesale power markets
Unregulated non-energy companies	Generation plants, refineries, and oil and gas production facilities
Cooperatives	Generation, transmission, and distribution
Foreign entities	Several utilities and power stations

Table 3.3. Average Emission Levels in the Production of 1 MWh of Electricity (pounds of emission per MWh)

	Coal	Oil	Natural Gas	Nuclear
Carbon dioxide	2,249	1,672	1,135	0
Sulfur dioxide	13	12	0.1	0
Nitrogen oxides	6	4	1.7	0

Source: EIA 2003.

In addition to dwindling supplies of oil, various problems with burning coal, and a limited supply of natural gas, all of these conventional sources of energy contribute to environmental pollution problems. Nuclear power, on the other hand, produces electricity through the fission of uranium, not the burning of fuels. Consequently, nuclear power plants do not pollute the air with nitrogen oxides, sulfur oxides, dust, or greenhouse gases like carbon dioxide. There is a stark difference between the emission levels of coal, oil, and natural gas as compared to nuclear (see table 3.3).

Beyond safety considerations involved with proper operation of a nuclear power plant, the huge drawback of using nuclear power is the production of nuclear waste, along with its safe transportation and storage or disposal.

KEY TERMS: NUCLEAR POWER

Before beginning a basic discussion of nuclear energy and nuclear power in the United States, it is important to define key terms (EIA 2002).

conventional mill (uranium)—A facility engineered and built principally for processing of uraniferous ore materials mined from the earth and the recovery, by chemical treatment in the mill's circuits, of uranium and/or other valued coproduct components from the processed ore.

cutoff grade—The lowest grade, in percentage of U3O8, of uranium ore at a minimum specified thickness that can be mined at specified cost.

development drilling—Drilling done to determine more precisely size, grade, and configuration of an ore deposit subsequent to the time the determination is made that the deposit can be commercially developed.

domestic—Domestic means within the fifty states, District of Columbia, Puerto Rico, the Virgin Islands, Guam, and other U.S. possessions. The word *domestic* is used also in conjunction with data and information that are compiled to characterize a particular segment or aspect of the uranium industry in the United States.

domestic uranium industry—Collectively, those businesses (whether U.S. or foreign based) that operate under the laws and regulations pertaining to the conduct of commerce within the United States and its territories and possessions and that engage in activities within the United States and its territories and possessions specifically directed toward uranium exploration, development, mining, and milling; marketing of uranium materials; enrichment; fabrication; or acquisition and management of uranium materials for use in commercial nuclear power plants.

exploration drilling—Drilling done in search of new mineral deposits, on extensions of known ore deposits, or at the location of a discovery up to the time when the company decides that sufficient ore reserves are present to justify commercial exploitation. Assessment drilling is reported as exploration drilling.

fabricated fuel—Fuel assemblies composed of an array of fuel rods loaded with pellets of enriched uranium dioxide.

foreign purchase—A uranium purchase of foreign-origin uranium from a firm located outside of the United States.

heap leach solutions—The separation, or dissolving out, from mined rock of the soluble uranium constituents by the natural action of percolating a prepared chemical solution through mounded (heaped) rock material. The mounded material usually contains low-grade mineralized material and/or waste rock produced from open pit or underground mines. The solutions are collected after percolation is competed and processed to recover the valued components.

in situ leach mining (ISL)—The recovery, by chemical leaching, of the valuable components of an ore body without physical extraction of the ore from the ground. Also referred to as "solution mining."

milling of uranium—The processing of uranium from ore mined by conventional methods, such as underground or open pit, to separate the uranium from the undesired material in the ore.

National Uranium Resource Evaluation (NURE)—A program begun by the U.S. Atomic Energy Commission (AEC) in 1974 to make a comprehensive evaluation of U.S. uranium resources and continued through 1983 by the AEC's successor agencies, the Energy Research and Development Administration (ERDA) and the Department of Energy (DOE). The NURE program included aerial radiometric and magnetic surveys, hydrogeochemical and stream boreholes, and geological studies to identify and evaluate geological environments favorable for uranium.

nonconventional plant (uranium)—A facility engineered and built principally for processing of uraniferous solutions that are produced during in situ leach mining, from heap leaching, or in the manufacture of other commodities, and the recovery, by chemical treatment in the plant's circuits, of uranium from the processed solutions.

nuclear electric power (nuclear power)—Electricity generated by an electric power plant whose turbines are driven by steam produced by the heat from fission of nuclear fuel in a reactor.

nuclear reactor—An apparatus in which a nuclear fission chain reaction can be initiated, controlled, and sustained at a specific rate. A reactor includes fuel (fissionable material), moderating material to control the rate of fission, a heavy-walled pressure vessel to house reactor components, shielding to protect personnel, a system to conduct heat away from the reactor, and instrumentation for monitoring and controlling the reactor's systems. In the vast majority of the world's nuclear power plants, the heat energy generated by uranium fuel is transferred to ordinary water and is carried away from the reactor's core either as steam in boiling-water reactors (BWRs) or as superheated water in pressurized-water reactors (PWRs).

- *Boiling-water reactor (BWR)*—In a typical commercial boiling-water reactor, (1) the reactor core creates heat, (2) a steam-water mixture is produced when very pure water (reactor coolant) moves upward through the core, absorbing heat, (3) the steam-water mixture leaves the top of the core and enters the two stages of moisture separation, where water droplets are removed before the steam is allowed to enter the steam line, and (4) the steam line directs the steam to the main turbine, causing it to turn the turbine generator, which produces electricity. The unused steam is exhausted to the condenser, where it is condensed into water. The resulting water is pumped out of the condenser with a series of pumps, reheated, and pumped back to the reactor vessel. The reactor's core contains fuel assemblies that are cooled by water, which is force-circulated by electrically powered pumps. Emergency cooling water is supplied by other pumps that can be powered by on-site diesel generators. Other safety systems, such as the containment cooling system, also need electric power.
- *Pressurized-water reactor (PWR) and reactor vessel*—In a typical commercial pressurized light-water reactor, (1) the reactor core generates heat, (2) pressurized water in the primary coolant loop carries the heat to the steam generator, (3) inside the steam generator heat from the primary coolant loop vaporizes the water in a secondary loop, producing steam, and (4) the steam line directs the steam to the main turbine, causing it to turn the turbine generator, which produces electricity. The unused steam is exhausted to the condenser, where it is condensed into water. The resulting water is pumped out of the condenser with a series of pumps, reheated, and pumped back to the steam generator. The reactor core contains fuel assemblies that are cooled by water, which is force-circulated by electrically powered pumps. Emergency cooling water is supplied by other pumps, which can be powered by on-site diesel generators. Other safety systems, such as the containment cooling system, also need power.

processing uranium—Uranium-recovery operations at a mill, in situ leach plant, by-product plant, or other type of recovery operation.

reclamation—Process of restoring surface environment to acceptable preexisting conditions. Includes surface contouring, equipment removal, well plugging, revegetation, and so on.

restoration—The returning of all affected groundwater to its premining quality for its premining use by employing the best practical technology.

separative work units (SWU)—The standard measure of enrichment services.

uranium—A heavy, naturally radioactive, metallic element (atomic number 92). Its two principally occurring isotopes are ^{235}U and ^{238}U. The isotope ^{235}U is indispensable to the nuclear industry because it is the only isotope existing in nature to any appreciable extent that is fissionable by thermal neutrons. The isotope ^{235}U is also important because it absorbs neutrons to produce a radioactive isotope that subsequently decays to the isotope ^{239}Pu, which also is fissionable by thermal neutrons.

uranium concentrate—A yellow or brown powder obtained by the milling of uranium ore, processing of in situ leach mining solutions, or as a by-product of phosphoric acid production.

uranium deposit—A discrete concentration of uranium mineralization that is of possible economic interest.

uranium endowment—The uranium that is estimated to occur in rock with a grade of at least 0.01 percent U3O8. The estimate of the uranium endowment is made before consideration of economic availability and any associated uranium resources.

uranium hexafluoride (UF6)—A white solid obtained by chemical treatment of U3O8 that forms a vapor at temperatures above 56° C. UF6 is the form of uranium required for the enrichment process.

uranium ore—Rock containing uranium mineralization in concentrations that can be mined economically (typically one to four pounds of U3O8 per ton, or 0.05 to 0.20 percent U3O8).

uranium oxide—Uranium concentrate or yellowcake. Abbreviated as U3O8.

uranium property—A specific tract of land with known uranium reserves that could be developed for mining.

uranium reserves—Estimated quantities of uranium in known mineral deposits of such size, grade, and configuration that the uranium could be recovered at or below a specified production cost with currently proven mining and processing technology and under current law and regulations. Reserves are based on direct radiometric and chemical measurements of drill hole and other types of sampling of the deposit. Mineral grades and thickness, spatial relationships, depths below the surface, mining and reclamation methods, distance to milling facilities, and amenability of ores to processing are considered in the evaluation. The amount of uranium in ore that could be exploited within the chosen forward-cost levels are estimated utilizing available sampling, engineering, geologic, and economic data in accordance with conventional engineering practices.

uranium resource categories—Three categories of uranium resources are used to reflect differing levels of confidence in the resources reported. Reasonably assured resources (RAR), estimated additional resources (EAR), and speculative resources (SR) are described below.

- *Reasonably assured resources (RAR)*—The uranium that occurs in known mineral deposits of such size, grade, and configuration that it could be recovered within the given production cost ranges, with currently proven mining and processing technology. Estimates of tonnage and grade are based on specific sample data and measurements of the deposits and on knowledge of deposit characteristics. RAR correspond to DOE's uranium reserves category.
- *Estimated additional resources (EAR)*—The uranium in addition to RAR that is expected to occur, mostly on the basis of direct geological evidence, in extensions of well-explored deposits, little-explored deposits, and undiscovered deposits believed to exist along well-defined geological trends with known deposits, such that the uranium can subsequently be recovered within the given cost ranges. Estimates of tonnage and grade are based on available sampling data and on knowledge of the deposit characteristics, as determined in the best-known parts of the deposit or in similar deposits. EAR correspond to DOE's probable potential resources category.
- *Speculative resources (SR)*—Uranium in addition to EAR that is thought to exist, mostly on the basis of indirect evidence and geological extrapolations, in deposits discoverable with existing extrapolation techniques. The locations of deposits in this category can generally be specified only as being somewhere within given regions or geological trends. The estimates in this category of SR correspond to DOE's possible resources plus speculative potential resources categories combined.

yellowcake—A natural uranium concentrate that takes its name from its color and texture. Yellowcake typically contains 70 to 90 percent U_3O_8 by weight. It is used as feedstock for uranium fuel enrichment and fuel pellet fabrication.

BASICS

Nuclear energy is energy from atoms. The energy is in the nucleus (core) of an atom. Atoms are tiny particles that make up every object in the universe. There is enormous energy in the bonds that hold atoms together.

Nuclear energy can be used to make electricity. But first the energy must be released. It can be released from atoms in two ways: nuclear fusion and nuclear fission. In nuclear fission, atoms are split apart to form smaller atoms, releasing energy. Nuclear power plants use this energy to produce electricity. In nuclear fusion, energy is released when atoms are combined or fused together to form a larger atom. This is how the sun produces energy. Fusion is the subject of ongoing research, but it is not yet clear that it will ever be a commercially viable technology for electricity generation.

DID YOU KNOW?

All nuclear power in the United States is used to generate electricity.

The fuel most widely used by nuclear plants for nuclear fission is uranium. Uranium is nonrenewable, though it is a common metal found in rocks all over the world. Nuclear fission uses a certain kind of uranium, referred to as U-235. This kind of uranium is used as fuel because its atoms are easily split apart. Though uranium is quite common, about a hundred times more common than silver, U-235 is relatively rare. Most U.S. uranium is mined in the western United States. Once uranium is mined, the U-235 must be extracted and processed before it can be used as a fuel.

DID YOU KNOW?

Almost 20–22 percent of the world's known reserves of uranium are located in Australia, which notably has no nuclear power plants.

During nuclear fission, a small particle called a neutron hits the uranium atom and splits it, releasing a great amount of energy as heat and radiation. More neutrons are also released. These neutrons go on to bombard other uranium atoms, and the process repeats itself over and over again. This is called a chain reaction (see figure 3.1).

U.S. NUCLEAR POWER: AN OVERVIEW (EIA 2009B)

Nuclear power plants rely on the heat energy generated from nuclear fission to provide the power for a turbo-generator. In this process, the nucleus of a heavy element, such as uranium or plutonium, splits when bombarded by a neutron in a nuclear reactor. The fission process for uranium atoms typically yields two smaller atoms called fission fragments, two or three neutrons, plus about two hundred million electron volts of nuclear energy in the form of radiation and heat (see figure 3.1). Because more neutrons are released from a uranium fission event than are required to initiate the event,

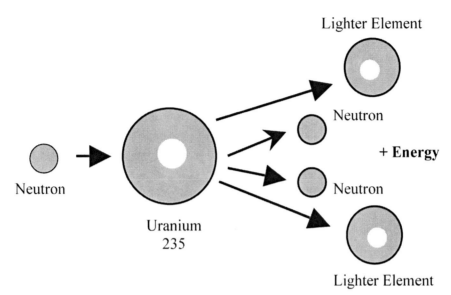

FIGURE 3.1
How Fission Splits the Uranium Atom

the reaction can become self-sustaining—a chain reaction—under controlled conditions, thus producing a tremendous amount of energy.

In many nuclear power plants, the heat energy generated by uranium fuel is transferred to ordinary water and is carried away from the reactor's core either as steam in boiling-water reactors (BWRs) or as superheated water in pressurized-water reactors (PWRs). In a PWR the superheated water in the primary cooling loop flows through a special heat exchanger called a "steam generator" that is used to boil water and create steam in a secondary loop that feeds the turbo-generator. The two-loop design of a PWR keeps the radioactivity isolated, and only clean steam is circulated through the turbine. This helps to minimize maintenance costs and radiation exposures to the plant personnel.

In the BWR reactor design, the water is boiled inside the reactor core and then sent directly to the turbo-generator to produce power. The BWR is a simpler design, but it has the disadvantage that the turbo-generator and more of the plant become slightly radioactive over time. In either a boiling-water or pressurized-water installation, steam under pressure turns a steam turbine, whose shaft is connected to an electrical generator.

The turbo-generator and many auxiliary systems are similar to conventional fossil fuel plants. The main difference is the nuclear heat source and the reactor safety systems. Boiling-water and pressurized-water reactors are sometimes called light-water reactors (LWRs) because they utilize ordinary water (H2O) to transfer the heat energy from reactor to turbine in the electricity generation process. In other reactor designs, the heat energy is transferred by heavy water (D2O), helium gas, or a liquid metal such as lead or sodium.

Because the cooling water absorbs some of the neutrons needed for the nuclear chain reaction and the natural uranium absorbs others, the concentration of the U-235 isotope in uranium (0.72 percent level of natural uranium) must be increased (enriched) above the level of natural uranium (3–4 percent low-enriched reactor grade) to assist in sustaining the nuclear chain reaction in the reactor core. The process of enrichment is explained later.

A light-water nuclear power reactor can have up to eight hundred fuel assemblies in its fuel core. An assembly consists of a group of sealed fuel rods, each filled with UO2 (uranium dioxide) pellets, held in place by end plates and supported by metal spacer grids to brace the rods and maintain the proper distances between them. The fuel core can be thought of as a reservoir from which heat energy can be extracted through the nuclear chain reaction process. During the operation of the reactor, the concentration of U-235 in the fuel is decreased as those atoms undergo nuclear fission to create heat energy. Some U-235 atoms are converted to atoms of fissile Pu-239, some of which will, in turn, undergo fission and produce energy. The products created by the nuclear fission reactions are retained within the fuel pellets, and these become neutron-absorbing products (called "poisons") that act to slow the rate of nuclear fission and heat production. As the reactor operation is continued, a point is reached at which the declining concentration of fissile nuclei in the fuel and the increasing concentration of poisons result in lower than optimal heat energy generation, and the reactor must be shut down temporarily and refueled. The fraction of the reactor's fuel core replaced during refueling is typically one-fourth for a boiling-water reactor and one-third for a pressurized-water reactor.

The amount of energy available in nuclear fuel can be expressed in "full-power days." The number of full-power days between refueling outages is related to the amount of fissile U-235 contained in the fuel assemblies at the beginning of the cycle, but it is also limited by safety and engineering considerations.

The amount of energy extracted from nuclear fuel is called "burn up," which is expressed in terms of the heat energy produced per initial unit of fuel weight. Burn up is commonly expressed as megawatt days thermal per metric ton of initial heavy metal. Less than 1 percent of the uranium is typically burned in a reactor before it is discarded as "spent fuel." This is because of the buildup of poisons in the fuel from the fission process and because the metal cladding on the fuel weakens over time as it is exposed to radiation. A higher percentage of U-235 in the core at the beginning of a cycle will permit the reactor to be run for a greater number of full-power days. It is also possible to recycle the spent fuel in LWRs to extract more energy from the remaining unused uranium. However, in the United States, the "once-through fuel cycle" is currently preferred over recycling.

The Nuclear Fuel Cycle

Figure 3.2 illustrates the nuclear fuel cycle for a typical light-water reactor. The cycle consists of "front-end" steps that lead to the preparation of uranium for use as fuel and "back-end" steps that are necessary to safely manage and dispose of the highly radioactive spent nuclear fuel. The front end of the nuclear fuel cycle commonly is separated into the following steps: exploration, mining, milling, conversion, enrichment, and fabrication.

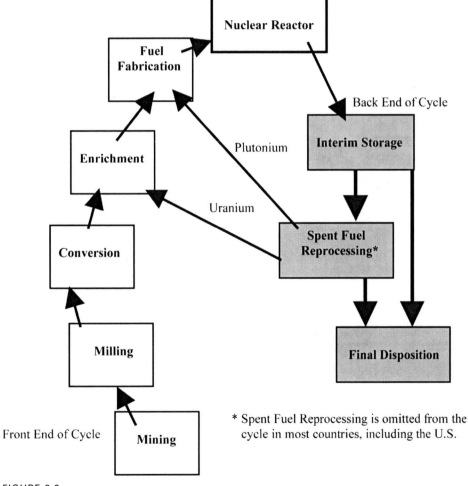

FIGURE 3.2
The Nuclear Fuel Cycle

Front End of Cycle

- **Exploration**—The nuclear fuel cycle begins with the exploration for uranium deposits. Many types of sophisticated techniques are used to find uranium, such as geophysical and geochemical analyses, satellite and airborne radiometric surveys, water sampling, and drilling. Because uranium deposits usually occur in discrete little pockets rather than long, continuous seams like coal and oil, the exploration process can be difficult and expensive. A drill hole, for example, can just miss a huge deposit, thereby giving a false indication that there is no uranium present. Likewise, the drill hole can also produce misleading results when it hits a tiny pocket of uranium. Therefore, many drill holes are usually needed to characterize the extent of the ore deposit. Once a uranium deposit has been discovered, geologists go to work to evaluate the extent and grade (concentration) of the ore deposit. Mining engineers then evaluate how best to extract the ore and what it will cost. This is combined to generate a "uranium reserves estimate" that is basically a table of how much uranium can be extracted at a particular cost. The reserves estimate can then be compared with the market price of uranium to determine whether it makes sense to open a mine.

- **Mining**—If the mining operation appears profitable, the next step in the fuel cycle is to begin mining uranium ore. Uranium ore is the raw rock or gravel material that comes from the ground. It can be extracted through conventional mining in open pit and underground methods similar to those used for mining other metals. Another popular mining technique is called solution mining or in situ leach mining. In this technology, a slightly caustic liquid such as carbonated water is injected underground to dissolve the uranium in place. The water containing dissolved uranium is then pumped back to the surface, where the uranium is then recovered. Some people liken this environmentally friendly mining technique to operating a simple water purification plant. Uranium ores in the United States typically range anywhere from about 0.05 to 0.3 percent uranium oxide (U_3O_8). Uranium is also present in some phosphate rocks. Although the uranium concentrations are quite low (fifty to two hundred parts per million) in the phosphate rock, the mining operations can be profitable because of the phosphates and phosphoric acid that is coproduced. At one time, phosphates were a significant source of uranium production in the United States.

- **Milling**—The next step in the nuclear fuel cycle, called milling, involves the purification and concentration of uranium. Mined uranium ores normally are processed by grinding the ore materials to a uniform particle size and then treating the ore with acid material to extract the uranium by chemical leaching. The mining process commonly yields a dry powdered-uranium oxide (U_3O_8) material called "yellowcake" because of its bright yellow color.

- **Uranium conversion**—Milled uranium oxide, U3O8, must be converted to uranium hexafluoride, UF6, in order to separate the individual U-235 atoms from the U-238 atoms. A solid at room temperature, UF6 can be changed to a gaseous form at moderately higher temperatures. The UF6 conversion product contains only natural, not enriched, uranium. Currently, there is only one conversion plant in the United States.

- **Enrichment**—The concentration of the fissionable isotope U-235 (0.71 percent in natural uranium) is less than that required to sustain a nuclear chain reaction in light-water reactor cores. Natural UF6 thus must be "enriched" in the fissionable isotope for it to be used as nuclear fuel. Light-water reactor fuel normally is enriched up to about 4 percent U-235. However, different levels of enrichment for a particular nuclear fuel application may be specified. The UF6 gas that remains has less than normal concentrations of U-235 in it and is therefore called "the enrichment tails," "depleted uranium," or simply "DU."

  Gaseous diffusion and gas centrifuge are the commonly used uranium enrichment technologies. The gaseous diffusion process consists of passing the natural UF6 gas feed under high pressure through a series of diffusion barriers (semiporous membranes) that permit passage of the lighter 235-UF6 atoms at a faster rate than the heavier 238-UF6 atoms. This differential treatment, applied across a large number of diffusion "stages," progressively raises the product stream concentration of U-235 relative to U-238. In the gaseous diffusion technology, the separation achieved per diffusion stage is relatively low, and a large number of stages is required to achieve the desired level of isotope enrichment. Because this technology requires a large capital outlay for facilities and consumes large amounts of electrical energy, it is relatively expensive.

  In the gas centrifuge process, the natural UF6 gas is spun at high speed in a series of cylinders. This acts to separate the 235-UF6 and 238-UF6 molecules based on their slightly different atomic masses. Gas centrifuge technology involves relatively high capital costs for the specialized equipment required, but its power costs are significantly lower than those for the gaseous diffusion technology. New enrichment technologies currently being developed are the atomic vapor laser isotope separation (ALVIS) and the molecular laser isotope separation (MLIS). Each laser-based enrichment process can achieve higher initial enrichment (isotope separation) factors than the diffusion or centrifuge processes can achieve.

- **Fabrication**—For use as nuclear fuel, enriched UF6 is next converted into uranium dioxide (UO2) powder, which is then processed into pellet form. The pellets are then fired in a high-temperature sintering furnace to create hard ceramic pellets of enriched uranium. The cylindrical pellets then undergo a grinding process to achieve a uniform pellet size. The pellets are stacked, according to each nuclear core's design specifications, into tubes of corrosion-resistant metal alloy. The tubes are sealed to contain the fuel pellets: these tubes are called fuel rods. The finished fuel rods are bundled together in special fuel assemblies that are then used to make up the nuclear fuel core of a power reactor.

Back End of Cycle

The back end of the cycle is divided into the following steps: interim storage, reprocessing, and waste disposal.

- **Interim storage**—After its operating cycle, the reactor is shut down for refueling. The fuel discharged at that time (spent fuel) is initially stored underwater at the reactor site in special leaktight pools with cooling systems. The cooling systems are needed because the spent fuel continues to emit radiation and give off heat even after it is removed from the reactor. This is called decay heat, and the levels can be as much as 5 percent of the original heat level. The spent fuel cools off, and after about five years it can be taken out of the water and placed in special dry storage canisters made of concrete for long-term storage. Reactor owners have been able to store the fuel onsite safely now for decades. However, some pools are beginning to fill up to capacity. At these sites, special dry storage pads are being installed to store the fuel in dry canisters until a final underground repository is constructed. These storage locations are known as independent spent fuel storage installations (ISFSI) at the reactor site or at a facility away from the site.

- **Reprocessing**—Spent fuel discharged from light-water reactors contains appreciable quantities of fissile (U-235, Pu-239), fertile (U-238), and other radioactive materials. These fissile and fertile materials can be chemically separated and recovered from the spent fuel. The recovered uranium and plutonium can, if economic and institutional conditions permit, be recycled for use as nuclear fuel. Currently, plants in Europe are reprocessing spent fuel from utilities in Europe and Japan. In the United States, DOE is conducting research on advanced nuclear fuel cycles that may someday recycle the spent fuel and extract more energy from the uranium. Advanced fuel cycle research is also being conducted to reduce the amount of nuclear waste that must be buried in the repository.

- **Waste disposal**—Although the safety record for spent fuel storage systems has been good, ultimately the waste stream must be isolated from the biosphere until the radioactivity contained in them has diminished to a safe level. Under the Nuclear Waste Policy Act of 1982, as amended, the Department of Energy has responsibility for the development of the waste disposal system for spent nuclear fuel and high-level radioactive waste. Current plans call for the ultimate disposal of the wastes in solid form in licensed deep, stable geologic structures. A national repository at Yucca

Mountain has been chosen to store the waste, but progress on the facility has been delayed. In the interim, spent fuel is being stored at about seventy sites across the United States.

NUCLEAR POWER SECTOR: A REGULATED ENVIRONMENT

Even before 9/11, the nuclear power sector was (and is) one of the nation's most regulated industries. It is subject to numerous environmental regulations as well as the voluntary obligations imposed by the NRC/DOE/EPA's environmental health and safety improvement initiatives. Federal/state OSHA statutes and major federal EPA statutes, as well as numerous state laws, impose significant compliance and reporting requirements on the sector (see sidebar 3.1).

Sidebar 3.1. Major Health, Safety, and Environmental Legislation Pertinent to the Nuclear Power Sector

1. **Public Health Service Act (1944) as amended in 1957, 1958, 1960, 1976 (42 USC 201 et seq.)** provides EPA with the authority to conduct a range of radiation protection activities:

   - Collect, analyze, and interpret data on environmental radiation levels.
   - Research the environmental and human health effects of exposure to radiation.
   - Develop protective action guides.
   - Provide training and technical assistance to the sites. It also provides the authority to develop and implement a plan to effectively provide personnel, equipment, medical supplies, or other federal resources in responding to health emergencies.

2. **1946 Atomic Energy Act as amended in 1954 (42 USC 2011 et seq.)** established the Atomic Energy Commission (AEC) to promote the "utilization of atomic energy for peaceful purposes to the maximum extent consistent with the common defense and security and with the health and safety of the public." When EPA was formed, the AEC's authority to issue generally applicable environmental radiation standards was transferred to EPA. Other federal and state organizations must follow these standards when developing requirements for their areas of radiation protection.

   EPA also received the Federal Radiation Council's authority under the AEA:

   - To develop guidance for federal and state agencies contain recommendations for their use in developing radiation protection requirements.
   - To work with states to establish and execute radiation protection programs.

3. **Energy Policy and Conservation Act of 1975** began the modern energy policy era, which, not coincidentally, was a response to an oil price spike after an OPEC

(continued)

embargo in 1973–1974. The act not only created the Strategic Petroleum Reserve to counter severe disruptions in the nation's oil supply but also introduced for the first time Corporate Average Fuel Economy (CAFE) standards for automobile manufacturers, requiring that average fuel economy of vehicles sold by auto manufacturers in the U.S. achieve twice the fuel efficiency of their then current levels.

4. **1963 Clean Air Act (CAA) as amended in 1970, 1977, 1990 (42 USC 7401 et seq.)** was first passed in 1955 as the Air Pollution Control, Research and Technical Assistance Act and amended in 1963 to become the CAA. A more significant statute was passed in 1970 and amended in 1977 and 1990. It provides EPA the authority to regulate air pollutants from a wide variety of sources, including automobiles, electric power plants, chemical plants, and other industrial sources. Section 112 provides EPA the authority to list hazardous air pollutants, or HAPs, and to develop and enforce emission limits for each of them. Section 112(a) introduced the concept of "ample margin of safety to protect public health" in setting these limits. The limits are referred to as "National Emission Standards for Hazardous air Pollutants," or NESHAPs. Section 103 of the CAA provides EPA the broad authority to gather information, provide grants, conduct or promote research, and coordinate and accelerate training.

5. **Occupational Safety and Health Act (OSH Act) of 1970** provides the Department of Labor authority to set comprehensive workplace safety and health standards, including permissible exposures to chemicals in the workplace, and authority to conduct inspections and issue citations for violations of safety and health regulations.

6. **Marine Protection, Research, and Sanctuaries Act as amended in 1977 (32 USC 1401 et seq.)** authorizes EPA to issue permits and promulgate regulations for disposing of materials into the territorial waters of the United States when it will not degrade or endanger human health, human welfare, ecological systems, the marine environment, or the economy.

7. **Federal Water Pollution Control Act amended by the Clean Water Act (CWA) of 1977, 1987 (33 USC 1251 et seq.)** was first enacted in 1948 as the Federal Water Pollution Control Act. Subsequent extensive amendments defined the statute to be known as the CWA in 1972; it was further amended in 1977 and 987. The CWA provides EPA authority to regulate effluents from sewage treatment works, refineries, chemical plants, and other industry sources into U.S. waterways. EPA has recently undertaken control efforts in on-point source pollution as well. The primary objective of the CWA is to restore and maintain the integrity of the nation's waters. The CWA sets these provisions:

   ▪ Performance standards for major industries to ensure pollution control

   ▪ Requirement for states and tribes to set specific water quality criteria appropriate for the waters and to develop pollution control programs for meeting them

   ▪ A permitting process to ensure that any development or other activity in valuable wetlands and other aquatic habitats is conducted in an environmentally sound manner

- Authority for EPA to initiate an emergency response to any accidental release of oil or hazardous substances, including radionuclides, when there is a substantial threat to the public health or welfare. This authority is similar to that assigned by the Comprehensive Environmental Response, Compensation and Liability Act. However, it applies only to discharges or potential discharges of pollutant to navigable waters of the U.S.

8. **1974 Safe Drinking Water Act (SWDA) as amended in 1977, 1986, 1996 (43 USC s/s 300f et seq.)** requires EPA to promulgate and enforce primary standards for contaminants in public water systems, including radionuclides. Initially, EPA was to set interim regulations for a limited group of contaminants and later revise those regulations to set standards for the remaining contaminants. The 1986 amendments required EPA to develop maximum contaminant level guidelines (MCLGs) and maximum contaminant levels, or MCLs, concurrently and to finalize the interim regulations. Under this statute EPA may delegate program enforcement authority to the states. The 1996 amendments to the SDWA directed EPA to do the following:

- Withdraw the Notice of Proposed Drinking Water Rule, including the proposed MCLGs and MCLs and monitoring, reporting, and public notification requirements for radon, due to the controversy over the cost-benefit basis for the proposed standard.

- Arrange for the National Academy of Sciences (NAS) to conduct a formal study of radon in drinking water; publish a risks reduction and cost analysis for possible radon MCLs by February 1999; promulgate the radon MCLG and National Primary Drinking Water Regulation (NPDWR) for drinking water by the year 2000; develop an alternative MCL for radon, as directed to ensure that any revised drinking water standard will maintain or increase public health protection; and review all drinking water regulations every six years.

9. **Resource Conservation and Recovery Act (RCRA) of 1976 as amended in 1984, 1986 (42 USC 6901 et seq.)** provides EPA with authority to establish standards and regulations for handling and disposing of solid and hazardous wastes (cradle-to-grave provisions). RCRA has four main goals:

- Protect human health and the environment from hazards posed by waste disposal.

- Conserve energy and natural resources through waste recycling and recovery.

- Reduce or eliminate the generation of waste, including hazardous waste.

- Ensure that wastes are managed in an environmentally safe manner.

RCRA focuses only on active and planned facilities. It does not address abandoned or historical sites. Although sources, special nuclear, or by-product material as defined by the AEA is specially excluded from RCRA, it generally does apply to naturally occurring radioactive materials.

10. **1978 Uranium Mill Tailings Radiation Control Act (UMTRCA) (4 USC 1022 et seq.)** amended the AEA by directing EPA to set generally applicable health and

(*continued*)

environmental standards to govern the stabilization, restoration, disposal, and control of effluents and emissions at both active and inactive mill tailings sites.

Title 1 of the act covers inactive uranium mill tailing sites, depository sites, and vicinity properties. It directs EPA, the Department of Energy, and the Nuclear Regulatory Commission to undertake the following:

EPA: Must set standards that provide protection as consistent with the requirements of RCRA as possible. The standards must include groundwater protection limits.

DOE: Implements EPA's standards for the tailings piles and the vicinity properties, and provides perpetual care for some properties.

NRC: Reviews completed site cleanups for compliance with EPA standards. Licenses the site to the state or DOE for perpetual care.

Title II of the act covers operating uranium processing sites licensed by the NRC. The EPA was directed to promulgate disposal standards in compliance with Subtitle C of the Solid Waste Disposal Act, as amended, to be implemented by NRC or the Agreement States. The 1993 amendments to UMTRCA further directed EPA to promulgate general environmental standards for the processing, possession, transfer, and disposal of uranium mill tailings. NRC was required to implement these standards at Title II sites.

11. **1980 Low-Level Radioactive Waste Policy Act (LLRWPA) as amended in 1985 (42 USC 2021b et. Seq.)** requires each state to be responsible for providing disposal capacity for commercial low-level waste (LLW) generated within its borders by January 1, 1986. It encouraged states to form regional compacts to develop new disposal facilities. The LLRWPA was amended in 1985 to provide states more time to develop facilities and to provide incentives for volume reduction of LLW.

12. **Comprehensive Environmental Response Compensation and Liability Act of 1980 (CERCLA) and the Superfund Amendments and Reauthorization Act of 1986 (SARA) (42 USC 9601 et seq.)** provides the basic legal framework for the federal "Superfund" program to clean up abandoned hazardous waste sites. It creates a tax on the chemical and petroleum industries and provides broad federal authority to respond directly to releases or threatened releases of hazardous substances pollutants and contaminants that may endanger public health or the environment. CERCLA also authorizes and directs EPA to carry out a program of training and evaluation of training needs in the procedures for the handling and removal of hazardous substances. CERCLA applies to hazardous substances defined by other environmental laws. For example, since the CAA amendments list radionuclides as hazardous substances, they are covered by CERCLA. CERCLA authorizes two kinds of response actions:

- *Short-term removal actions* address actual or threatened releases requiring prompt response to protect human health or the environment at any site.

- *Long-term remedial actions* address actual or threats of releases that are serious but not immediately life threatening. EPA conducts long-term remedial response actions only at sites on the National Priority List, commonly known as Superfund sites.

These response actions are conducted in accordance with the concept of operations contained in the NCP (40 CFR Part 300).

While the above is a general description of how CERCLA works, response actions are handled somewhat differently at federal and federally licensed sites. This reflects the division of radiation protection responsibilities among various federal agencies. For example:

- The Department of Defense and the Department of Energy coordinate response actions at their sites.

- Nuclear Regulatory Commission regulations—rather than CERCLA regulations—govern response actions at some NRC-licensed facilities (Nuclear Regulatory Commission–licensed facilities that are covered by the Price Anderson Amendments [Section 170] of the AEA).

13. **1982 Nuclear Waste Policy Act (NWPA) (42 USC 10101 et seq.)** supports the use of deep geologic repositories for the safe storage and/or disposal of radioactive waste. The act establishes procedures to evaluate and select sites for geologic repositories and for the interaction of state and federal governments. It also provides a timetable of key milestones the federal agencies must meet in carrying out the program.

The NWPA assigns DOE the responsibility to site, build, and operate a deep geologic repository for the disposal of HLW and spent nuclear fuel (SNF). It directs EPA to develop standards for protection of the general environment from offsite releases of radioactive material in repositories. The act directs NRC to license DOE to operate a repository only if it meets EPA's standards and all other relevant requirements.

14. **Emergency Planning and Community Right-to-Know Act of 1986,** also known as SARA Title III, mandates state and community development of emergency preparedness plans and also establishes an annual manufacturing-sector emissions reporting program.

15. **Hazardous Materials Transportation Act (HMTA)** provides the Department of Transportation the authority to regulate the packaging and movement of hazardous materials.

16. **1987 Nuclear Waste Policy Amendments Act (NWPAA) (42 USC 10101 et seq.)** directs DOE to consider Yucca Mountain as the primary site for the first geological repository for HLW and SNF and prohibits DOE from conducting site-specific activities at a second site unless authorized by Congress. It also requires the secretary of energy to develop a report on the need for a second repository no later than January 1, 2010. The NNWPAA also established a commission to study the need and feasibility of a monitored retrievable storage facility.

(continued)

17. **1988 Indoor Radon Abatement Act (IRAA)** establishes a long-term goal that indoor air be as free from radon as the ambient air outside buildings. The law authorized funds for radon-related activities at the state and federal levels:

    - establishing state programs and providing technical assistance
    - conducting radon surveys of schools and federal buildings
    - establishing training centers and a proficiency program for firms offering radon services
    - developing model construction standards
    - developing a citizen's guide to radon

18. **Pollution Prevention of Act of 1990** makes it the national policy of the United States to reduce or eliminate the generation of waste at the source whenever feasible and directs the EPA to undertake a multimedia program of information collection, technology transfer, and financial assistance to the states to implement this policy and to promote the use of source waste reduction techniques.

19. **Energy Policy Act (EnPA) of 1992** introduced one of the key drivers of the renewable energy industry to date, the Production Tax Credit (PTC), which offers to independent power producers a subsidy per kilowatt hour generated from renewable sources for a period of ten years from the beginning of power generation. The EnPA requires EPA to "promulgate standards to ensure protection of public health from high-level radioactive wastes in a deep geologic repository that might be built under Yucca Mountain in Nevada." It directs EPA to issue these site-specific public health and safety standards, "based upon and consistent with the findings and recommendations of the National Academy of Sciences. . . ."

20. **1992 Waste Isolation Pilot Plant Land Withdrawal Act (WIPP LWA) as amended in 1996 (L 102-579)** sets aside the land for developing and building a transuranic radioactive waste repository and assigns the following specific regularity and enforcement roles to EPA:

    - Finalize generally applicable standards for disposal of SNF, HLW, and TRU (transuranic). (These standards limit releases of radioactive materials to the environment and notify future generations of the location and content of the disposal facility.)
    - Develop criteria for the disposal of TRU at the WIPP that are consistent with the generally applicable standards.
    - Certify that the WIPP complies with the generally applicable standards if DOE satisfies the criteria.
    - Ensure that the WIPP complies with other applicable environmental and public health and safety regulations.
    - Reevaluate the WIPP every five years to determine whether it should be recertified.

It also prohibits the application of the general standards to sites being considered for the disposal of HLW and SNF under the NWPA (e.g., Yucca Mountain).

The 1996 WIPP LWA Amendments (PL104-201) did the following:

- Specified November 30, 1997, as a nonbinding date for the WIPP site to open, pending certification by EPA that the site meets environmental regulatory requirements

- Exempted the WIPP from RCRA Land Disposal Requirements. It withdrew requirements in the original act that DOE conduct underground tests on-site with transuranic waste to determine whether it could be disposed of safely.

21. **Federal Energy Regulatory Commission (FERC)** is an independent regulatory agency within the Department of Energy and is perhaps the leading regulatory body in determining the impact on the average consumer of energy. FERC has jurisdiction over electricity pricing, licensing for hydroelectric plants and liquid natural gas (LNG) terminals, and oil pipeline transport rates.

22. **Energy Policy Act of 2005** does everything from extending daylight saving time to authorizing fifty million dollars in grants for biomass energy projects to changing the depreciation allowances for gas distribution lines.

23. **State regulations**. State governments are increasingly active in the environmental and safety areas.

24. **Energy Reorganization Act of 1974** established the Nuclear Regulatory Commission. The act split the responsibility of the U.S. AEC that was established under the AEA into two separate agencies. Before this law the U.S. AEC maintained complete control over both the development and production of nuclear weapons and civilian use of radioactive material and the regulation of such agencies. Under the ERA the DOE was given control over the production of nuclear weapons and other uses of nuclear power. The Nuclear Regulatory Commission was given the task of the regulatory work.

25. **Nuclear Regulatory Commission regulations**. Laws established by the NRC as to radiation safety can be found in Title 10 of the Code of Federal Regulations. Some notable sections of Title 10 are as follows:

10 CFR 19—**Training requirements**. Contains the requirements for training radiation worker as well as inspection requirements for licensed nuclear facilities.

10 CFR 20—**Standards for protection against radiation**. Standards established under the AEA and the ERA. It establishes standards for the protection against ionizing radiation occurring at facilities operating under an NRC license. The purpose of this standard is to minimize the dose received during the receipt, possession, use, transfer, and disposal of licensed material so as to not exceed the standards prescribed in this regulation.

(continued)

10 CFR 51—**Environmental protection regulation for domestic licensing.**
Standard established governing all applicable environmental protection
regulations at facilities licensed by the NRC.

10 CFR 61—**Licensing requirements for land disposal of radioactive waste.**
Determines the terms and conditions for license issuing regarding the
procedure and criteria for land disposal of radioactive waste.

10 CFR 61—**Packing and transportation of radioactive material.** Determines
requirements for packaging, preparation for shipment, and transportation
of NRC licensed material. This section also dictates the standards for NRC
approval of packaging and shipping of material.

DID YOU KNOW?

The costs of meeting mandated and self-imposed environmental and security requirements are high and continue to rise. Indeed, an increasing
portion of new P&E investment is for environmental/security improvement purposes rather than to improve productivity or increase output.

THE BOTTOM LINE

Pre-9/11—Technical innovations and developments in digital information and
telecommunications dramatically increased interdependencies among the nation's
critical infrastructures. Each infrastructure depends on other infrastructures to
function successfully. Disruption in a single infrastructure can generate disturbances within other infrastructures and over long distances, and the pattern of
interconnections can extend or amplify the effects of a disruption. The energy (including nuclear) infrastructure provides essential fuel to all of the other critical infrastructures and in turn depends on the nation's transportation, communications,
finance, and government infrastructures. For example, coal shipments are highly
dependent on rail. There are also interdependencies with the energy infrastructure
itself, particularly the dependence of petroleum refineries and pipeline pumping
stations on a reliable electricity supply, backup generators, and utility maintenance
vehicles to be supplied with diesel and gasoline fuel.

Post-9/11—The tragic events of 9/11 were a wake-up call for all of us. Like never
before in the modern era, the terrorists focused our attention like a laser beam on
security of the homeland and in particular our critical infrastructure. History shows

that over the years terrorists have used energy infrastructure as primary targets, causing governments to take extraordinary prevention actions.

REFERENCES AND RECOMMENDED READING

Bond, C. 2002. Statement on S. 2579. *Congressional Record*, daily edition.

DHS. 2003. *The national strategy for the physical protection of critical infrastructure and key assets.* Washington, DC.

DHS. 2008. Homeland security presidential directives. www.dhs.gov/xabout/laws/editorial_0607.shtm.

DHS. 2007. *Energy: Critical infrastructure and key resources sector-specific plan as input to the national infrastructure protection plan.* Washington, D.C.

DOL. 2008. *Communication.* Washington, DC: U.S. Department of Labor.

EIA (Energy Information Administration). 2002. *Nuclear glossary.* www.eia.doe.gov/cneaf/nuclear/page/glossary.html (accessed October 18, 2009).

EIA. 2003. *Carbon dioxide emissions from generation of electric power.* www.eia.doe.gov/cneaf/electricity.html (accessed October 19, 2009).

EIA. 2009a. *Official energy statistics.* www.eia.doe.gov/fuelelectric.html (accessed April 26, 2009).

EIA. 2009b. *Introduction to nuclear power.* http://www.eia.doe.gov/cneaf/nuclear/page/intro.html (accessed October 21, 2009).

USDOE. 2009. *Science and technology.* http://www.energy.gov/sciencetech/index.htm (accessed April 25, 2009).

U.S. International Trade Commission. 1991. *Global competitiveness of U.S. advanced-technology manufacturing industries: Pharmaceuticals.* Report to the Committee of Finance, U.S. Senate, on Investigation No. 332-302 under Section 332(g) of the Tariff Act of 1930.

United States Nuclear Regulatory Commission. *NRC regulations title 10, code of federal regulations.* http://www.nrc.gov/reading-rm/doc-collections/cfr/ (accessed November 8, 2009).

Nuclear Power Sector Security

The emergence of amorphous and largely unknown terrorist individuals and groups operating independently (freelancers) and the new recruitment patterns of some groups, such as recruiting suicide commandos, female and child terrorists, and scientists capable of developing weapons of mass destruction, provide a measure of urgency to increasing our understanding of the psychological and sociological dynamics of terrorist groups and individuals (Hudson 1999).

   The only truly secure system is one that is powered off, cast in a block of concrete and sealed in a lead-lined room with armed guards.

—*Gene Spafford*

NRC (2009) points out that U.S. nuclear power plants are some of the most fortified civilian facilities in the country. After 9/11, the NRC used its independent regulatory authority to order the nuclear industry to implement new defensive capabilities, more rigorous guard training, and many other security enhancements. In response, the industry has met the increased requirements regardless of cost. The process of upgrading security continues.

INTRODUCTION

While it is true that the U.S. nuclear power sector has not been attacked on a large scale by terrorists, it is also true that the potential risk is more than conjecture. From several terrorism incidents on oil and gas pipelines in South America, the Middle East, and Africa, it is apparent that energy infrastructure, including nuclear power, is a preferred target of the terrorists. Moreover, al Qaeda has made it clear that it has an interest in attacking American nuclear power plants, nuclear waste storage areas, and associated nuclear power infrastructure.

The obvious question might be: What makes the nuclear power sector components attractive targets for terrorists? It is an attractive target because

- Such an attack has the potential to inflict devastating damage—huge bang (or scare factor) for the buck.
- A 2004 study by the Union of Concerned Scientists says a successful attack on the Indian Point nuclear facility thirty-five miles north of Manhattan could cause as many as 44,000 near-term casualties and 500,000 long-term deaths from cancer.
- An attack on a nuclear power plant could release a high level of radiation that would gravely endanger public health.
- Nuclear power sector components, assets, and infrastructure support facilities (e.g., waste storage sites) are spread throughout the nation with little definition of boundaries.
- Much of the nuclear power sector equipment is one-of-a-kind large machinery that is difficult to replace in the short term.
- Several nuclear sector storage components were initially designed and constructed without concern for terrorist intrusion or destructive activities.
- Many nuclear power systems are monitored and operated using underprotected SCADA systems (Kaplan 2006).

Even though it was first considered an environmental issue (e.g., nuclear radiation spilling from some nuclear source and contaminating the environmental media: air, water, and/or soil), nuclear power sector safety and security has been an issue of congressional interest for many years. Although there have been no credible attacks against any of the U.S. nuclear power facilities or nuclear waste sites, review of the historical incidents listed in sidebar 4.1 provides some perspective on the types of targets and magnitude of the consequences that could result from terrorist attacks on energy sector components. Remember, the primary purpose of U.S. nuclear power plants is to provide electricity (energy). Therefore, even though the incidents listed in sidebar 4.1 are not directly related to nuclear power production or nuclear waste facilities, the listed incidents demonstrate the lengths to which criminals and terrorists will go. Obviously, these incidents (and others) increase concern over the potential for shutdown of our economy, damage to or shutdown of interrelated critical infrastructure, and the associated potential safety and health impact on surrounding areas. Thus, it is important to ensure that the person(s) designated as site safety/security professional(s) be fully trained in the profession, experienced, and properly certified.

Sidebar 4.1. MU.S. Energy Sector Terrorism/Criminal Incidents*

Event 1: Copper-wire theft causes power disruptions (May 20, 2006)—Thieves broke into all four substations of Bangor Hydro-Electric C, breaking locks and tampering with wiring, making the electrical system unsafe in addition to the cable theft. Power loss was experienced throughout the area for about six hours.

Event 2: Attempted copper-wire theft (March 29, 2005)—Someone broke into the Omaha Public Power District substation and cut two 13,800-volt power lines. One of the power lines was a backup circuit for part of downtown Omaha's power. Omaha police believe the person who cut the lines is seriously burned. When power employees arrived on the scene, they found a saw and smelled burned flesh. Authorities believe the person who cut the line may have been one of the homeless persons who frequently visit the area of the substation and may have been trying to steal the lines for the copper that's inside.

Event 3: Drug addicts stealing power lines to feed their methamphetamine habits (March 16, 2005)—In Portland, Oregon, police arrested four people with six hundred to eight hundred pounds of stolen power lines. Detectives say the suspects took the line from a Portland General Electric (PGE) substation and planned to sell it for scrap, using the money for methamphetamine. PGE says the value of the copper used to make power lines has recently skyrocketed, so thieves are desperate to steal it. Some even risk their lives by trying to swipe live lines. It is illegal for scrap yards to knowingly accept stolen metal. However, there is no law requiring yards to ask where the seller acquired the metal.

Event 4: Criminals hack into power grid (March 11, 2005)—Hackers caused no serious damage to Washington, DC, systems that feed the nation's power grid, but their untiring efforts have heightened concerns that electric companies have failed to adequately fortify defenses against a potential catastrophic strike. The fear is that in a worst-case scenario, terrorists or others could engineer an attack that sets off a widespread blackout and damages power plants, prolonging an outage. The Federal Energy Regulatory Commission and others in the industry said the companies' computer security is uneven. The biggest threat to the grid, analysts said, may come from power companies using older equipment that is more susceptible to attack.

Event 5: Greenpeace activists climb smokestack (February 15, 2005)—Six Greenpeace activists were sentenced to jail terms ranging from five to thirty days for climbing a smokestack at a coal-fire power plant in protest of President Bush's energy policy. The protesters cut a hole in the fence that surrounds Allegheny

(continued)

---

* The information is a consolidated listing of events, activities, and news accounts as reported by the U.S. Department of Homeland Security and other government agencies. Many of these events were reported by various newspapers throughout the U.S.

Energy's Hartfield Ferry Power Station, then climbed the 700-foot smokestack and unfurled a 2,500-square-foot banner. The six pleaded guilty to misdemeanor charges of reckless endangerment, failure to disperse, disorderly conduct, and defiant trespass.

Event 6: Electricity thefts (January 30, 2005)—Rising energy prices, tighter competition, and the desire to keep bystanders from being electrocuted have led utilities nationwide to fight back with growing fervor. They're hiring full-time investigative teams and running nighttime surveillance. Energy thieves cost U.S. utilities between $4 billion and $6 billion yearly. To stem the losses, the nation's investor-owned utility companies hire an average of ten full-time investigators each to root out theft, according to the Edison Electric Institute, a trade association for U.S. shareholder-owned electric companies. Utilities prosecute 10 percent of cases.

Event 7: Gas-pipe valve vandalized (January 6, 2005)—In Kearney, Nebraska, vandals turned off a gas-pipeline valve to about 150 homes and businesses in frigid temperatures.

Event 8: Electrical power lines sabotaged (December 28, 2004)—In Nevada, eight high-voltage transmission lines servicing the northern region and Reno area were sabotaged. Officials reported that the collapse of any one tower could possibly bring down a string of other towers.

Event 9: Attempted bombing of power substation (December 16, 2004)—Utility workers investigating a power outage found cut fencing and an unexploded homemade bomb in a Northern Indiana Public Service Company substation.

Event 10: Vandalism of transmission towers (October 14, 2004)—Four men were reported taking photos and shooting videos outside the headquarters of a Wisconsin transmission company. At the same time, two of the transmission towers collapsed after the bolts were removed.

Event 11: Fake bomb found at electrical tower (October 11, 2004)—A fake bomb at a Philadelphia electrical tower forced the closure of a major highway connecting Philadelphia with its western suburbs for several hours. A box made to look like an explosive but later determined to be a hoax was discovered at the foot of an electric transmission tower. An FBI spokesperson did not know whether the incident was related to the sabotage of an electrical-transmission tower in Wisconsin on October 10.

Event 12: Pipeline explosion (September 28, 2004)—Police in New Caney, Texas, connected a pipeline explosion to an August vandalism case. Evidence pointed to one or more suspects behind the explosion and the ensuing six-hour fire. No one was injured, and nearby houses and other structures were not affected. Damage to construction equipment, electricity-transmission lines, and the six-inch pipeline itself were estimated to be over one million dollars. Evidence showed that a track hoe and a bucket truck at the explosion site were tampered with.

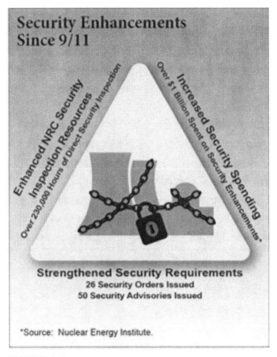

FIGURE 4.1
Security Enhancements

NRC SECURITY MEASURES

The NRC requires that nuclear power plants be both safe and secure. Safety refers to operating the plant in a manner that protects the public and the environment. Security refers to protection inside the plant—using people, equipment, and fortifications—from intruders who wish to damage or destroy it in order to harm people and the environment. The process of upgrading safety and security is ongoing, never ending, and incident responsive—the 9/11 tragedy being an incident example. Security enhancements at nuclear power plants since 9/11 are shown in figure 4.1.

LAYERS OF DEFENSE*

U.S. NRC has developed a security plan that is characterized by a layered defense scheme. The layered security scheme consists primarily of protecting against aircraft, defending against adversaries, preparedness and response, and security materials.

---

* Material in this section is from NRC's *Security spotlight.* http://www.nrc.gov/reading-rm/doc-collections/fact-sheets/security-spotlight/aircraft.html (accessed October 30, 2009).

Protecting against Aircraft

Since 9/11, the NRC has comprehensively studied the effect of an air attack on nuclear power plants. Shortly after 9/11, the NRC began a security and engineering review of operating nuclear power plants. Assisting the NRC were national experts from Department of Energy laboratories, who used state-of-the-art experiments and structural and fire analyses.

Even though these classified studies confirm that there is a low likelihood that an air attack on a nuclear power plant would affect public health and safety (thanks in part to the inherent robustness of the structures), the NRC is now considering new regulations for further reactors' security.

It is the federal government and military's responsibility to protect the nation against an aircraft attack. To that end, the NRC works closely with its federal partners to identify and implement enhanced security programs, including

- Military and Department of Homeland Security program to identify and protect critical infrastructure
- Criminal history checks on flight crews
- Reinforced cockpit doors
- Checking of passenger list against "no-fly" lists
- Increased control of cargo
- Random inspections
- Increased federal air marshal presence
- Improved screening of passengers and baggage
- Controls on foreign passenger carriers
- Improved coordination and communication between civilian and military authorities

Defending against Adversaries

Commercial nuclear power plants are heavily fortified with well-trained and armed guards. They also have layered physical security measures, such as access control, water barriers, intrusion detection, and strategically placed guard towers. Together, these make up the plants' response to the design basis threat—usually called the DBT. The DBT is developed from real-world intelligence information and describes the adversary force—coming from both ground and water—the plants must defend against. The two licensed category I fuel-cycle facilities in the U.S. that make reactor fuel for nuclear plants must also define a DBT similar to that of nuclear power plants. DBT specifics are not public in order to protect sensitive information that could aid terrorists. The NRC regularly reviews the DBT and adds new requirements when necessary.

The NRC routinely tests the security at nuclear facilities with realistic exercises using a well-trained mock adversary force. These force-on-force exercises are designed to test a security force's ability to defend against the DBT. The NRC oversees every aspect of these exercises and evaluates them using rigorous standards. These exercises typically span several days. During the attack, the mock adversary force tires to reach and damage key safety systems. Any significant security problems are promptly identified, reviewed, and fixed prior to NRC's inspection team leaving the facility. The NRC tests every plant with a force-on-force exercise a minimum of every three years. The plants also must conduct their own yearly exercises.

Preparedness and Response

No matter how small the risk, the NRC requires all nuclear power plants to have and periodically test emergency plans that are coordinated with federal, state, and local responders. The goal of preparedness is to reduce the risk to the public during an emergency.

In an emergency, the NRC and the licensee would activate their incident response program. Licensee specialists would evaluate the situation and identify ways to end the emergency, while the NRC would monitor the event closely, keeping government offices informed. If a radiation release occurred, the plant would make protective action recommendations to state and local officials, such as evacuating areas around the plant.

Emergency Planning Zones (EPZs)

Each nuclear power plant has two emergency planning zones (EPZs). Each EPZ considers the specific conditions and geography at the site and in the surrounding community. The first is the plume exposure pathway EPZ, which has a radius of about ten miles from the reactor. People living there may be asked to evacuate or "shelter in place" during an emergency to avoid or reduce their radiation dose. The second is the ingestion exposure pathway EPZ. This has a radius of about fifty miles from the reactor to protect residents in the area from consuming contaminated food and water.

Response Modes

The NRC uses these modes for responding to events:

- Monitoring—A heightened state of readiness for getting and accessing incident information.
- Activation—A team of reactor and preparedness specialists begin staffing the headquarters operations center and regional incident response centers to respond to the event. Another team of specialists travels to the site if needed.

Securing Materials

Radioactive materials are used in many beneficial ways, including medical (cancer treatment), academic, and industrial uses. Despite these benefits, some materials can potentially harm people and the environment if misused. For these reasons, their security, including use and handling, is strictly regulated in the United States by the NRC.

- *Dirty bombs*—A "dirty bomb," also called a "radiological dispersal device" (RDD), combines explosives, such as dynamite, with radioactive material. A dirty bomb is *not* a nuclear weapon. Most dirty bombs would not be highly destructive and would not release enough radiation to kill people or cause severe illness. Instead, a dirty bomb is a "weapon of mass disruption" that could cause panic and fear and require costly cleanup. Some materials licensed by the NRC could possibly be used in a dirty bomb, which is why they are strictly regulated.
- *National Source Tracking System (NSTS)*—The NSTS is major security initiative of the NRC. This highly secure, accessible, and easy-to-use computer system tracks high-risk radioactive sources from the time they are manufactured through their import to the time of their disposal or export or until they decay enough to no longer be of concern. This website had been developed to provide stakeholders with the current status of program activities, tools, and resources to aid in using the system, and mechanisms to provide direct feedback to NRC.

HOMELAND SECURITY DIRECTIVES

As a result of 9/11, the Department of Homeland Security was formed. On matters pertaining to homeland security, homeland security presidential directives are issued by the president. Each directive has specific meaning and purpose and is carried out by the U.S. Department of Homeland Security. Each directive is listed and summarized in sidebar 4.2.

Sidebar 4.2.   Homeland Security Presidential Directives

HSPD-1: Organization and Operation of the Homeland Security Council. (White House) Ensures coordination of all homeland security-related activities among executive departments and agencies and promotes the effective development and implementation of all homeland security policies.

HSPD-2: Combating Terrorism through Immigration Policies. (White House) Provides for the creation of a task force that will work aggressively to prevent aliens who engage in or support terrorist activity from entering the United States and to detain, prosecute, or deport any such aliens who are within the United States.

HSPD-3: Homeland Security Advisory System. (White House) Establishes a comprehensive and effective means to disseminate information regarding the risk of terrorist acts to federal, state, and local authorities and to the American people.

HSPD-4: National Strategy to Combat Weapons of Mass Destruction. Applies new technologies and increased emphasis on intelligence collection and analysis, strengthens alliance relationships, and establishes new partnerships with former adversaries to counter this threat in all of its dimensions.

HSPD-5: Management of Domestic Incidents. (White House) Enhances the ability of the United States to manage domestic incidents by establishing a single, comprehensive national incident management system.

HSPD-6: Integration and Use of Screening Information. (White House) Provides for the establishment of the Terrorist Threat Integration Center.

HSPD-7: Critical Infrastructure Identification, Prioritization, and Protection. (White House) Establishes a national policy for federal departments and agencies to identify and prioritize United States critical infrastructure and key resources and to protect them from terrorist attacks.

HSPD-8: National Preparedness. (White House) Identifies steps for improved coordination in response to incidents. This directive describes the way federal departments and agencies will prepare for such a response, including prevention activities during the early stages of a terrorism incident. This directive is a companion to HSPD-5.

HSPD-8, Annex 1: National Planning. Further enhances the preparedness of the United States by formally establishing a standard and comprehensive approach to national planning.

HSPD-9: Defense of United States Agriculture and Food. (White House) Establishes a national policy to defend the agriculture and food system against terrorist attacks, major disasters, and other emergencies.

HSPD-10: Biodefense for the Twenty-First Century. (White House) Provides a comprehensive framework for our nation's biodefense.

HSPD-11: Comprehensive Terrorist-Related Screening Procedures. (White House) Implements a coordinated and comprehensive approach to terrorist-related screening that supports homeland security at home and abroad. This directive builds on HSPD-6.

HSPD-12: Policy for a Common Identification Standard for Federal Employees and Contractors. (White House) Establishes a mandatory government-wide standard for secure and reliable forms of identification issued by the federal government to its employees and contractors (including contractor employees).

HSPD-13: Maritime Security Policy. Establishes policy guidelines to enhance national and homeland security by protecting U.S. maritime interests.

HSPD-16: Aviation Strategy. Details a strategic vision on aviation security while recognizing ongoing efforts, and directs the production of a national strategy for aviation security and supporting plans.

(*continued*)

HSPD-18: Medical Countermeasures against Weapons of Mass Destruction. (White House) Establishes policy guidelines to draw upon the considerable potential of the scientific community in the public and private sectors to address medical countermeasure requirements relating to CBRN threats.

HSPD-19: Combating Terrorist Use of Explosives in the United States. (White House) Establishes a national policy, and calls for the development of a national strategy and implementation plan, on the prevention and detection of, protection against, and response to terrorist use of explosives in the United States.

HSPD-20: National Continuity Policy. (White House) Establishes a comprehensive national policy on the continuity of federal government structures and operations and a single National Continuity Coordinator responsible for coordinating the development and implementation of federal continuity policies.

HSPD-21: Public Health and Medical Preparedness. (White House) Establishes a national strategy that will enable a level of public health and medical preparedness sufficient to address a range of possible disasters.

Source: DHS (2009).

REFERENCES AND RECOMMENDED READING

CBS. 2006. *The explosion at Texas City.* www.cbsnews.com/stories/2006 /10/26/60minutes/ main2126509.shtml (accessed April 27, 2008).

Crayton, J. W. 1983. Terrorism and the psychology of the self. In *Perspectives on Terrorism,* ed. Lawrence Zelic Freedman and Yonah Alexander, 33–41. Wilmington, DE: Scholarly Resources.

DHS. 2009. *Homeland security presidential directives.* http://www.dhs.gov/xabout/laws/ editorial_0607.htm (accessed October 31, 2009).

Erikson, E. 1994. *Identity of the life cycle.* New York: W. W. Norton.

Ferracuti, F. 1982. A sociopsychiatric interpretation of terrorism. *Annals of the American Academy of Political and Social Science* 463 (September), 129–41.

Fields, R. M. 1979. Child terror victims and adult terrorists. *Journal of Psychohistory* 7(1).

Federal Register. 2007. *Federal Register* 17688-17745.

Gurr, T. R. 1971. *Why men rebel.* Princeton, NJ: Princeton University Press.

HighBeam. 2008. *Explosion at First Chemical Corporation plant.* www.highbeam.com/ doc/1G1-92783471.html (accessed April 28, 2008).

HSE. 2008a. *Release of hydrofluoric acid from Marathon Petroleum Refinery, Texas, USA.* www.hse.gov.uk/comah/sragtech/casemarathon87.htm (accessed April 27, 2008).

HSE. 2008b. *PEMEX LPG Terminal, Mexico City, Mexico.* www.hse.gov.uk/comah/sragtech/casepemex84.htm (accessed April 27, 2008).

Hudson, R. A. 1999. *The sociology and psychology of terrorism: Who becomes a terrorist and why?* Washington, DC: Library of Congress.

Kaplan, E. 2006. *Anti-terror measures at U.S. nuclear plants.* http://www.cfr.org/publications/10450/antiterror_measures_at_us_nuclear_plants.html (accessed October 29, 2009).

Lees, Frank. 1996. *Loss prevention in the process industries.* 3:A5.1–A5.11. New York: Butterworth-Heinemann.

Local 1259. 2008. *The Texas City disaster.* www.local1259iaff.org/disaster.html (accessed April 27, 2008).

Long, D. E. 1990. *The anatomy of terrorism.* New York: Free Press.

Lyman, E. S. 2004. *Impacts of a terrorist attack at Indian Point Nuclear Power Plant.* http://www.ucsusa.org/nuclear_power/nuclear_power_risk/sabotage_and_attacks_on_reactors/impacts-of-a-terrorist-attack.html (accessed October 29, 2009).

Margolin, J. 1977. Psychological perspectives on terrorism. In *Terrorism: Interdisciplinary perspectives,* eds. Y. Alexander and S. M. Finger. New York: John Jay Press.

NRC. 2009. *Security spotlight.* http://www.nrc.gov/reading-rm/doc-collections/fact-sheets/security-spotlight/overview.html (accessed October 28, 2009).

Olson, M. 1971. *The logic of collective action.* Boston: Harvard University Press.

OMB. 1998. *Federal conformity assessment activities,* Circular A-119. Washington, DC: White House.

Pearlstein, R. 1991. *The mind of the political terrorist.* Wilmington, DE: Scholarly Resources.

RAE. 2008. Motiva Enterprises settles suit resulting from explosion at Delaware City refinery. www.caprep.com/0905038.htm (accessed April 27, 2008).

Sullivant, J. 2007. *Strategies for protecting national critical infrastructure assets: A focus on problem-solving.* New York: John Wiley & Sons.

U.S. EPA. 2008. Ashland oil spill, available at http://www.epa.gov/reg3/PA/ashlandoil/.

U.S. FA. 1989. *Phillips Petroleum chemical plant explosion and fire; Pasadena, Texas.* U.S. Fire Administration-TR-035. Emmitsburg, MD.

Wilkinson, P. 1974. *Political terrorism.* London: Macmillan.

Vulnerability Assessment (VA)

In God we trust, all others we monitor.

—*Intercept operator's motto*

Vulnerability means different things to different people . . . many associate vulnerability with a specific set of human activities . . . (Foster 1987).

INTRODUCTION*

One consequence of the events of 9/11 was the Department of Homeland Security's (DHS) directive to establish a critical infrastructure protection task force to ensure that activities to protect and secure vital infrastructure are comprehensive and carried out expeditiously. The perceived increase in security threats to critical infrastructure in the U.S. has led to numerous changes to facility security processes within many industries and municipalities.

Another consequence is a heightened concern among citizens in the United States over the security of their nuclear power infrastructure (i.e., the uninterrupted supply of electrical power and containment of any incident-released radiation). As mentioned, along with other critical infrastructure, the nuclear power sector is classified as a "vulnerable target" in the sense that inherent weaknesses in its operating environment could be exploited to cause harm to the system. There is also the possibility of a cascading effect—a chain of events—due to a terrorist act (foreign or domestic) affecting a nuclear power energy sector provider, which causes corresponding damage (collateral damage) to other nearby users. In addition to significant damage to the nation's energy sector, entities using nuclear power–derived energy (electricity)

---

* The information in this section is from or based on information in *Probabilistic Consequence Analysis of Security Threats—A Prototype Vulnerability Assessment Process for Nuclear Power Plants*, U.S. Department of Energy, Washington, DC; and ERRI, Palo Alto, CA (2004).

to produce finished products or to provide critical life-saving services (e.g., medical centers and hospitals) that come under terrorist attack could result in loss of life, shutdown of industry, catastrophic radioactive exposure and environmental damage to rivers, lakes, and wetlands, contamination of drinking water supplies, other short-term (instant death) and long-term public health impacts (cancer and death), undermining of public confidence in government, and disruption to commerce, the economy, and our normal way of life.

VULNERABILITIES

For the purpose of this text and according to FEMA (2008), vulnerability is defined as any weakness that can be exploited by an aggressor to make an asset susceptible to hazard damage.

In addition, according to the Department of Homeland Security (DHS 2009), vulnerabilities are physical features or operational attributes that render an entity open to exploitation or susceptible to a given hazard. Vulnerabilities may be associated with physical (e.g., a broken fence), cyber (e.g., lack of a firewall), or human (e.g., untrained guards) factors.

The FBI (2009) reports that since 9/11 there have been a variety of threats suggesting that U.S. energy facilities are being targeted for terrorist attacks. Although the information often is fragmentary and offers little insight into the timing and mode of an attack, the October 2002 operation against the French supertanker *Limburg* suggests that al Qaeda is serious about hitting the energy sector and its support structure.

In the U.S., al Qaeda appears to believe that an attack on oil and gas structures could do great damage to the U.S. economy. Moreover, terrorist planners probably perceive infrastructure such as dams and power lines as having softer defenses than other facilities.

According to DOE/EPRI (2004), significant security threats have previously been assessed by all nuclear power plants through the design basis threat evaluation process mandated by the NRC. Federal regulation 10 CFR 73, *Physical Protection of Plants and Materials* (see chapter 9), specifies the NRC's requirements for physical security at nuclear power plants and defines specific threats that each nuclear power plant must be prepared to defeat.

Although the security preparedness of nuclear power plants is notably more significant than that of other commercial industrial facilities, nuclear power facilities have also made significant security modifications since 9/11. Sites have increased security staffing, modified security equipment, and reevaluated response plans. The nuclear power industry has launched multiple assessment projects through industry associations. The NRC has issued several different orders to all commercial nuclear plants to implement compensatory security measures.*

It is important to point out that stemming from the increasingly recognized perspective that nuclear power plants have excelled in the protection of health and safety compared with other commercial industrial activities, the question of additional benefit from future security modifications is growing more significant.

VULNERABILITY ASSESSMENT (VA)[†]

A *vulnerability assessment* involves an in-depth analysis of the facility's functions, systems, and site characteristics to identify facility weaknesses and lack of redundancy and to determine mitigations or corrective actions that can be designed and implemented to reduce the vulnerabilities. A vulnerability assessment can be a stand-alone process or part of a full risk assessment. During this assessment, the analysis of site assets is based on (a) the identified threat, (b) the criticality of the assets, and (c) the level of protection chosen (i.e., based on willingness or unwillingness to accept risk).

It is important to point out that post-9/11, all sectors have taken great strides to protect their critical infrastructure. For instance, government and industry have developed vulnerability assessment methodologies for critical infrastructure systems and trained thousands of auditors and others to conduct them. The Public Health Security and Bioterrorism Preparedness and Response Act of 2002 requires community water systems serving more than 3,300 people to conduct a VA regarding terrorist attacks. A variety of security-related VA tools has been developed for water system administrators and is publicly available. Two examples are[‡]:

- *Protecting Your Community's Assets: A Guide for Small Wastewater Systems* (www .nesc.wvu.edu/netcsc)
- *Security Vulnerability Self-Assessment Guide for Small Drinking Water Systems Serving Populations of 3,300 and 10,000* (www.asdwa.org)

Portions of the chemical industry have voluntarily initiated VAs. For example, the American Chemistry Council requires its members to conduct VAs and implement

---

* NRC security orders can be found on the NCR website: http://www.nrc.gov/reading-rm/doc-collections/ enforcement/security/.

† Much of the information in this section is from U.S. Department of Energy (2002), *Vulnerability Assessment Methodology: Electric Power Infrastructure*, Washington, DC.

‡ Note that both of the these tools were used as guides in compiling *Water Infrastructure Protection and Homeland Security* (2007), Government Institutes Press.

security improvements. Examples of VA tools being used in the chemical industry include*

- *Guidelines for Analyzing and Managing the Security Vulnerabilities of Fixed Chemical Sites* (www.aiche.org/SVA/)
- *Site Security Guidelines for the U.S. Chemical Industry* (www.americanchemistry.com)
- *A Method to Assess the Vulnerability of U.S. Chemical Facilities* (www.ncjs.org)

The energy sector has also responded with general VA methodologies and tools. Three examples developed by the U.S. Department of Energy are[†]

- *Vulnerability and Risk Analysis Program: Overview of Assessment Methodology* (www esisac.com)
- *Energy Infrastructure Risk Management Checklists for Small and Medium Sized Energy Facilities* (www.esisac.com)
- *Vulnerability Assessment Methodology: Electric Power Infrastructure* (www.esisac.com)

The actual complexity of vulnerability assessments will range based upon the design and operation of the critical infrastructure system. The nature and extent of the VA will differ among systems based on a number of factors, including system size and potential population. Safety evaluations also vary based on knowledge and types of threats, available security technologies, and applicable local, state, and federal regulations. Preferably, a VA is "performance based," meaning that it evaluates the risk to the nuclear power plants based on the effectiveness (performance) of existing and planned measures to counteract adversarial actions.

With regard to nuclear power plants, the VA process provides a methodology that can be applied to a wide range of targets, plant configurations, and threat types.

The targets addressed in the VA include

- Reactor and its fuel load
- Spent fuel pool (SFP) and its fuel load

The plant configuration and operating modes addressed in the VA include:

- At-power configuration and operating modes
- Refuel configuration and operating modes

---

* Listed publications were the tools used to format *Chemical Infrastructure Protection and Homeland Security* (2009), Government Institutes Press.

† Listed publications were the tools used to format *Energy Infrastructure Protection and Homeland Security* (2010), Government Institutes Press.

In addition to the target, plant configurations, and threat types specifically supported in the VA, the methodology can be readily adapted for other nuclear power plant applications, such as

- Dry cask storage
- Spent fuel movement
- New threat types

When the process methodology listed above is combined with the following U.S. EPA (2002) requirements, it is made even more effective in guarding critical infrastructure. The common elements of USEPA vulnerability assessments are as follows:

- Characterization of the nuclear power sector, including its mission and objectives
- Identification and prioritization of adverse consequences to avoid
- Determination of critical assets that might be subject to malevolent acts that could result in undesired consequences
- Assessment of the likelihood (qualitative probability) of such malevolent acts from adversaries
- Evaluation of existing countermeasures
- Analysis of current risk and development of a prioritized plan for risk reduction

Benefits of Assessments

Nuclear power plant sector members should routinely perform vulnerability assessments to better understand threats and vulnerabilities, determine acceptable levels of risk, and stimulate action to mitigate identified vulnerabilities. The direct benefits of performing a vulnerability assessment include

- **Build and broaden awareness**—The assessment process directs senior management's attention to security. Security issues, risks, vulnerabilities, mitigation options, and best practices are brought to the surface. Awareness is one of the least expensive and most effective methods for improving the organization's overall security posture.
- **Establish or evaluate against a baseline**—If a baseline has been previously established, an assessment is an opportunity for a checkup to gauge the improvement or deterioration of a plant's security posture. If no previous baseline has been established (or the work was not uniform or comprehensive), an assessment is an opportunity to integrate and unify previous efforts, define common metrics, and establish a definitive baseline. The baseline also can be compared to best practices to provide perspective on a plant's security posture.

- **Identify vulnerabilities and develop responses**—Generating lists of vulnerabilities and potential responses is usually a core activity and outcome of an assessment. Sometimes, due to budget, time, complexity, and risk considerations, the response selected for many of the vulnerabilities may be inaction, but after completing the assessment process these decisions will be conscious ones, with a documented decision process and item-by-item rationale available for revisiting issues at scheduled intervals. This information can help drive or motivate the development of a risk management process.

- **Categorize key assets and drive the risk management process**—An assessment can be a vehicle for reaching corporate-wide consensus on a hierarchy of key assets. This ranking, combined with threat, vulnerability, and risk analysis, is at the heart of any risk management process. For many plants, Y2K was the first time a company-wide inventory and ranking of key assets was attempted. An assessment allows any plant to revisit that list from a broader and more comprehensive perspective.

- **Develop and build internal skills and expertise**—A security assessment, when not implemented in an audit mode, can serve as an excellent opportunity to build security skills and expertise within an organization. A well-structured assessment can have elements that serve as a forum for cross-cutting groups to come together and share issues, experiences, and expertise. External assessors can be instructed to emphasize "teaching and collaborating" rather than "evaluating" (the traditional role). Whatever a plant's current level of sophistication, a long-term goal should be to move that plant toward a capability for self-assessment.

- **Promote action**—Although disparate security efforts may be underway in an organization, an assessment can crystallize and focus management attention and resources on solving specific and systemic security problems. Often the people in the trenches are well aware of security issues (and even potential solutions) but are unable to convert their awareness to action. An assessment provides an outlet for their concerns and the potential to surface these issues at appropriate levels (legal, financial, executive) and achieve action. A well-designed and executed assessment not only identifies vulnerabilities and makes recommendations, it also gains executive buy-ins, identifies key players, and establishes a set of cross-cutting groups that can convert those recommendations into action.

- **Kick off an ongoing security effort**—An assessment can be used as a catalyst to involve people throughout the plant in security issues, build cross-cutting teams, establish permanent forums and councils, and harness the momentum generated by the assessment to build an ongoing institutional security effort. The assessment can lead to the creation of either an actual or a virtual (matrixed) security organization.

To facilitate the application of the VA process at a particular site, representative characteristics are provided. Most threat evaluations focus on a threat force gaining physical access to critical plant equipment within the plant-protected area. As a complement to those investigations and analyses, the VA includes examinations of plant threats that can be achieved from outside the plant-protected area. The threat characterizations provided include

- Insider sabotage (that is, tools in lieu of weapons)
- Small arms (handgun, assault rifle)
- Grenade launcher (rifle mounted)
- Sniper/anti-materiel rifle
- Medium and heavy machine gun
- Rocket-propelled grenade launcher (RPG-7)
- Mortars
- Explosive charges
- Incendiary weapons (e.g., grenades, rockets)
- Vehicle explosions
- Waterborne explosions
- Cyber attack
- Chemical agents

Note that these threats can be combined. For instance, most threats could be augmented by insider sabotage. The nuclear power plant VA process presented here differs from other industries' VA initiatives in the following ways:

- Detailed, representative threat characterizations are provided. Other industries' VAs generally require the assessment team to develop their own threat spectrum and characterizations.
- Nuclear power plant recommended threat characterizations include substantial weaponry threats. Other industries' VAs tend to focus on general security issues that nuclear power plants have already addressed through the design basis threat process.
- The nuclear power plant VA uses a quantitative probabilistic approach to develop an order-of-magnitude risk estimate for each threat-asset scenario, taking advantage of existing probabilistic risk assessments (PRAs). This provides a basis for screening a significant number of threat scenarios that are low risk. At the same time, it can identify some scenarios that do not screen and could be reduced in frequency by additional security precautions.

GENERAL SECTOR VULNERABILITY ASSESSMENT PROCESS

To demonstrate the differences and similarities of vulnerability assessments used in general-sector industries (i.e., water, wastewater, energy, chemical, agriculture, etc.) as

Table 5.1. Basic Elements in Vulnerability Assessments

Element	Points to Consider
1. Characterization of the entity, including its mission and objectives	• What are the important missions of the system to be assessed? Define the highest-priority services provided by the utility. Identify the industry's customers: 　° General public 　° Government 　° Military 　° Industrial 　° Critical care 　° Retail operations 　° Firefighting • What are the most important facilities, processes, and assets of the system for achieving the mission objectives and avoiding undesired consequences? Describe the: 　° Industry facilities 　° Operating procedures 　° Management practices that are necessary to achieve the mission objectives 　° How the industry operates 　° Treatment processes 　° Storage methods and capacity 　° Energy use and storage 　° Distribution system In assessing those assets that are critical, consider critical customers, dependence on other infrastructures (e.g., chemical, transportation, communications), contractual obligations, single points of failure, chemical hazards and other aspects of the industry's operations, or availability of industry utilities that may increase or decrease the criticality of specific facilities, processes and assets.
2. Identification and prioritization of adverse consequences to avoid	• Take into account the impacts that could substantially disrupt the ability of the system to provide a safe and reliable supply of energy. Sector systems should use the vulnerability assessment process to determine how to reduce risk associated with the consequences of significant concern. • Ranges of consequences or impacts for each of these events should be identified and defined. Factors to be considered in assessing the consequences may include 　° Magnitude of service disruption 　° Economic impact (such as replacement and installation costs for damaged critical assets or loss of revenue due to service outage) 　° Number of illnesses or deaths resulting from an event 　° Impact on public confidence in supply of service 　° Chronic problems arising from specific events 　° Other indicators of the impact of each event as determined by the particular sector. Risk reduction recommendations at the conclusion of the vulnerability assessment strive to prevent or reduce each of these consequences.
3. Determination of critical assets that might be subject to malevolent acts that could result in undesired consequences	• What are the malevolent acts that could reasonably cause undesired consequences? 　° Electronic, computer or other automated systems that are utilized by the energy sector entities (e.g., Supervisory Control and Data Acquisition (SCADA)

Element	Points to Consider
	◦ The use, storage, or handling of various energy materials (oil, coal, gas, etc.)
	◦ The operation and maintenance of such systems
4. Assessment of the likelihood (qualitative probability) of such malevolent acts from adversaries (e.g., terrorists, vandals)	• Determine the possible modes of attack that might result in consequences of significant concern based on critical assets of the particular sector entity. The objective of this step of the assessment is to move beyond what is merely possible and determine the likelihood of a particular attack scenario. This is a very difficult task as there is often insufficient information to determine the likelihood of a particular event with any degree of certainty.
	• The threats (the kind of adversary and the mode of attack) selected for consideration during a vulnerability assessment will dictate, to a great extent, the risk reduction measures that should be designed to counter the threat(s). Some vulnerability assessment methodologies refer to this as a "design basis threat" (DBT), where the threat serves as the basis for the design of countermeasures as well as the benchmark against which vulnerabilities are assessed. It should be noted that there is no single DBT or threat profile for all chemical systems in the United States. Differences in geographic location, size of the utility, previous attacks in the local area, and many other factors will influence the threat(s) that the energy sector entity should consider in its assessments. Specific sector entities should consult with the local FBI and/or other law enforcement agencies, public officials, and others to determine the threats upon which their risk reduction measures should be based.
5. Evaluation of existing countermeasures (Depending on countermeasures already in place, some critical assets may already be sufficiently protect. This step will aid in identification of the areas of greatest concern and help to focus priorities for risk reduction.)	• What capabilities does the system currently employ for detection, delay, and response?
	◦ Identify and evaluate current detection capabilities such as intrusion detection systems, energy quality monitoring, operational alarms, guard post orders, and employee security awareness programs.
	◦ Identify current delay mechanisms such as lock and key control, fencing, structure integrity of critical assets, and vehicle access checkpoints.
	◦ Identify existing policies and procedures for evaluation and response to intrusion and system malfunction alarms and cyber system intrusions.
	It is important to determine the performance characteristics. Poorly operated and maintained security technologies provide little or no protection.
	• What cyber protection system features does the facility have in place? Assess what protective measures are in place for the SCADA and business-related computer information systems such as
	◦ Firewalls
	◦ Modem access
	◦ Internet and other external connections, including wireless data and voice communications.
	◦ Security policies and protocols
	It is important to identify whether vendors have access rights and/or "backdoors" to conduct system diagnostics remotely.

(continued)

Table 5.1. (*continued*)

Element	Points to Consider

	• What security policies and procedures exist, and what is the compliance record for them? Identify existing policies and procedures concerning:
	  ○ Personal security
	  ○ Physical security
	  ○ Key and access badge control
	  ○ Control of system configuration and operational data
	  ○ Chemical and other vendor deliveries
	  ○ Security training and exercise records

6. Analysis of current risk and development of a prioritized plan for risk reduction	• Information gathered on threat, critical assets, energy sector operations, consequences, and existing countermeasures should be analyzed to determine the current level of risk. The utility should then determine whether current risks are acceptable or risk reduction measures should be pursued.
	• Recommended actions should measurably reduce risks by reducing vulnerabilities and/or consequences through improved deterrence, delay, detection, and/or response capabilities or by improving operational policies or procedures. Selection of specific risk reduction actions should be completed prior to considering the cost of the recommended action(s). Facilities should carefully consider both short- and long-term solutions. An analysis of the cost of short- and long-term risk reduction actions may impact which actions the utility chooses to achieve its security goals.
	• Facilities may also want to consider security improvements. Security and general infrastructure may provide significant multiple benefits. For example, improved treatment processes or system redundancies can both reduce vulnerabilities and enhance day-to-day operation.
	• Generally, strategies for reducing vulnerabilities fall into three broad categories:
	  ○ Sound business practices—affect policies, procedures, and training to improve the overall security-related culture at the facility. For example, it is important to ensure rapid communication capabilities exist between public health authorities and local law enforcement and emergency responders.
	  ○ System upgrades—include changes in operations, equipment, processes, or infrastructure itself that make the system fundamentally safer.
	  ○ Security upgrades—improve capabilities for detection, delay, or response.

compared to the nuclear power plant sector, table 5.1 provides an overview of the elements included in the general-sector assessment methodology. The elements included in this overview are based on actual in-field experience and lessons learned.

In table 5.1, step 3 deals with identification of asset criticality. This is an important first step in any vulnerability assessment. Identifying asset criticality serves several functions:

- It enables more careful consideration of factors that affect risk, including threats, vulnerabilities, and consequences of loss or compromise of the asset.

- It enables more focused and thorough consideration of loss or compromise of the asset.
- It enables leaders to develop robust methods for managing consequences of asset loss (restoration).
- It provides a means to increase awareness of a broad range of employees to protect truly critical assets and to differentiate in policies and procedures the heightened protection they require.

As previously indicated, identifying the criticality of assets is used primarily to focus the vulnerability analysis efforts. It also assists with the ranking of various recommendations for reducing vulnerabilities. Let's take a look at what electric power infrastructure critical assets might include,

Physical
- Generators
- Substations
- Transformers
- Transmission lines
- Distribution lines
- Control center
- Warehouses
- Office buildings
- Internal and external infrastructure dependencies

Cyber
- SCADA systems
- Networks
- Databases
- Business systems
- Telecommunications

Interdependencies
- Single-point nodes of failures
- Critical infrastructure components of high reliance

GENERAL SECTOR VULNERABILITY ASSESSMENT METHODOLOGY (U.S. DOE 2002)
Vulnerability assessment methodology consists of ten elements. Each element along with a description is listed below.

1. Network architecture
2. Threat environment
3. Penetration testing
4. Physical security
5. Physical asset analysis
6. Operations security
7. Policies and procedures
8. Impact analysis
9. Infrastructure interdependencies
10. Risk characterization

Network Architecture

This element provides an analysis of the information assurance features of the information network(s) associated with the organization's critical information systems. Information examined should include network topology and connectivity (including subnets), principal information assets, interface and communication protocols, function and linage of major software and hardware components (especially those associated with information security, such as intrusion detectors), and policies and procedures that govern security features of the network.

Procedures for information assurance in the system, including authentication of access and management of access authorization, should be reviewed. The assessment should identify any obvious concerns related to architectural vulnerabilities as well as operating procedures. Existing security plans should be evaluated, and the results of any prior testing should be analyzed. Results from the network architecture assessment should include potential recommendations for changes in the information architecture, functional areas, and categories where testing is needed, and suggestions regarding system design that would enable more effective information and information system protection.

Three techniques are often used in conducting the network architecture assessment:

1. Analysis of network and system documentation during and after the site visit
2. Interview with facility staff, managers, and chief information officer
3. Tours and physical inspections of key facilities

Threat Environment

Development of a clear understanding of the threat environment is a fundamental element of risk management. When combined with an appreciation of the value of the information assets and systems and the impact of unauthorized access and subsequent malicious activity, an understanding of threats provides a basis for better defining the level of investment needed to prevent such access.

The threat of a terrorist attack to the energy sector infrastructure is real and could come from several areas, including physical, cyber, and interdependency. In addition, threats could come from individuals or organizations motivated by financial gain or persons who derive pleasure from such penetration (e.g., recreational hackers, disgruntled employees). Other possible sources of threats are those who want to accomplish extremist goals (e.g., environmental terrorists, antinuclear advocates) or embarrass one or more organizations.

This element should include a characterization of these and other threats, identification of trends in these threats, and ways in which vulnerabilities are exploited. To the extent possible, characterization of the threat environment should be localized, that is, within the organization's service area.

Penetration Testing

The purpose of network penetration testing is to utilize active scanning and penetration tools to identity vulnerabilities that a determined adversary could easily exploit. Penetration testing can be customized to meet the specific needs and concerns of the energy sector unit or utility. In general, penetration testing should include a test plan and details on the rules of engagement (ROE). It should also include a general characterization of the access point of the critical information systems and include a general characterization of the access points to the critical information systems and communication interface connections, modem network connections, access points to principal network routers, and other external connections. Finally, penetration testing should include identified vulnerabilities and, in particular, whether access could be gained to the control network or specific subsystem or devices that have a critical role in assuring continuity of service.

Penetration testing consists of an overall process of establishing the ground rules or ROE for the test, establishing a "white cell" for continuous communication, developing a format or methodology for the test, conducting the test, and generating a final report that details methods, findings, and recommendations.

Penetration testing methodology consists of three phases: reconnaissance, scenario development, and exploitation. A one-time penetration test can provide the utility with valuable feedback; however, it is far more effective if performed on a regular basis. Repeated testing is recommended because new threats develop continuously, and the networks, computers, and architecture of the energy sector unit or utility are likely to change over time.

Physical Security

The purpose of physical security assessment is to examine and evaluate the systems in place (or being planned) and to identify potential improvements in this area for the sites evaluated. Physical security systems include access controls, barriers' locks and keys, badges and passes, intrusion detection devices and associated alarm reporting and display, closed-circuit television (assessment and surveillance), communications equipment (telephone, two-way radio, intercom, cellular), lighting (interior and exterior), power sources (line, battery, generator), inventory control, postings (signs), security system wiring, and protective force (see chapter 9 for greater detail on security devices). Physical security systems are reviewed for design, installation operation, maintenance, and testing.

The physical security assessment should focus on those sites directly related to the critical facilities, including information systems and assets required for operation. Typically included are facilities that house critical equipment or information assets or networks dedicated to the operation of electric, oil, or gas transmission, storage, or delivery systems. Other facilities can be included on the basis of criteria specified by the organization being assessed. Appropriate levels of physical security are contingent upon the value of company assets, the potential threats to these assets, and the cost associated with protecting the assets. Once the cost of implementing/maintaining a physical security program is known, it can be compared to the value of the company assets, thus providing the necessary information for risk management decisions. The focus of the physical security assessment task is determined by prioritizing the company assets, that is, the most critical assets receive the majority of the assessment activity.

At the start of the assessment, survey personnel should develop a prioritized listing of company assets. This list should be discussed with company personnel to identify areas of security strengths and weaknesses. During these initial interviews, assessment areas that would provide the most benefit to the company should be identified; once known, they should become the major focus of the assessment activities.

The physical security assessment of each focus area usually consists of the following:

- Physical security program (general)
- Physical security program (planning)
- Barriers
- Access controls/badges
- Locks/keys
- Intrusion detection systems
- Communications equipment
- Protective force/local law enforcement agency

The key to reviewing the above topics is not to merely identify whether they exist but to determine the appropriate level that is necessary and consistent with the value of the asset being protected. The physical security assessment worksheets provide guidance on appropriate levels of protection.

Once the focus and content of the assessment task have been identified, the approach to conducting the assessment can be either at the "implementation level" or at the "organizational level." The approach taken depends on the maturity of the security program.

For example, a company with a solid security infrastructure (staffing plans/procedures, funding) should receive a cursory review of these items; however, facilities where the security programs are being implemented should receive a detailed review. The security staff can act upon deficiencies found at the facilities, once reported.

For companies with an insufficient security organization, the majority of time spent on the assessment should take place at the organizational level to identify the appropriate staffing/funding necessary to implement security programs to protect company assets. Research into specific facility deficiencies should be limited to finding just enough examples to support any staffing/funding recommendations.

Physical Asset Analysis

The purpose of the physical asset analysis is to examine the systems and physical operational assets to ascertain whether vulnerabilities exist. Included in this element is an examination of asset utilization, system redundancies, and emergency operating procedures. Consideration should also be given to the topology and operating practices for electric and gas transmission, processing, storage, and delivery, looking specifically for those elements that either singly or in concert with other factors provide a high potential for disrupting service. This portion of the assessment determines company and industry trends regarding these physical assets. Historic trends, such as asset utilization, maintenance, new infrastructure investments, spare parts, SCADA linkages, and field personnel are part of the scoping element.

The proposed methodology for physical assets is based on a macro-level approach. The analysis can be performed with company data, public data, or both. Some companies might not have readily available data or might be reluctant to share that data.

Key output from analysis should be graphs that show trends. The historic data analysis should be supplemented with on-site interviews and visits. Items to focus on during a site visit include the following:

- Trends in field testing
- Trends in maintenance expenditures
- Trends in infrastructure investments
- Historic infrastructure outages
- Critical system components and potential system bottlenecks
- Overall system operation controls
- Use of and dependency on SCADA systems

- Linkages of operation staff with physical and IT security
- Adequate policies and procedures
- Communications with other regional utilities
- Communications with external infrastructure providers
- Adequate organizational structure

Operations Security

Operations security (OPSEC) is the systematic process of denying potential adversaries (including competitors or their agents) information about capabilities and intentions of the host organization. OPSEC involves identifying, controlling, and protecting generally nonsensitive activities concerning planning and execution of sensitive activities. The OPSEC assessment reviews the processes and practices employed for denying adversary access to sensitive and nonsensitive information that might inappropriately aid or abet an individual's or organization's disproportionate influence over system operation (e.g., electric markets, grid operations). This assessment should include a review of security training and awareness programs, discussions with key staff, and tours of appropriate principal facilities. Information that might be available through public access should also be reviewed.

Policies and Procedures

The policies and procedures by which security is administered (1) provide the basis for identifying and resolving issues, (2) establish the standards of reference for policy implementation, and (3) define and communicate roles, responsibilities, authorities, and accountabilities for all individuals and organizations interfaced with critical systems. They are the backbone for decisions and day-to-day security operations. Security policies and procedures become particularly important at times when multiple parties must interact to effect a desired level of security and when substantial legal ramifications could result from policy violations. Policies and procedures should be reviewed to determine whether they (1) address the key factors affecting security, (2) enable effective compliance, implementation, and enforcement, (3) reference or conform to established standards, (4) provide clear and comprehensive guidance, and (5) effectively address the roles, responsibilities, accountabilities, and authorities.

The objective of the policies and procedures assessment task is to develop a comprehensive understanding of how a facility protects its critical assets through the development and implementation of policies and procedures. Understanding and assessing this area provide a means of identifying strengths and areas for improvements that can be achieved through

- Modification of current policies and procedures
- Implementation of current polices and procedures
- Development and implementation of new policies and procedures
- Assurance of compliance with policies and procedures
- Cancellation of policies and procedures that are no longer relevant or are inappropriate for the facility's current strategy and operations

Impact Analysis

A detailed analysis should be conducted to determine the influence that exploitation of unauthorized access to critical facilities or information systems might have on an organization's operations (e.g., market and/or physical operations). In general, such an analysis would require a thorough understanding of (1) the applications and their information processing, (2) decisions influenced by this information, (3) independent checks and balances that might exist regarding information upon which decisions are made, (4) factors that might mitigate the impact of unauthorized access, and (5) secondary impacts of such access (e.g., potential destabilization of organizations serving the grid, particularly those affecting reliability or safety). Similarly, the physical chain of events following disruption, including the primary, secondary, and tertiary impacts of disruption, should be examined.

The purpose of the impact analysis is to help estimate the impact that outages could have on a specific sector unit. For example, outages in electric power, natural gas, and oil can have significant financial and external consequences to an energy sector unit. The impact analysis provides an introduction to risk characterization by providing quantitative estimates of these impacts so that the energy sector unit can implement a risk management program and weigh the risks and costs of various mitigation measures.

Infrastructure Interdependencies

The term *infrastructure interdependencies* refers to the physical and electronic (cyber) linkages within and among our nation's critical infrastructures—energy (electric power, oil, natural gas), telecommunications, transportation, water supply systems, banking and finance, emergency services, and government services. This task identifies the direct infrastructure linkages between and among the infrastructures that support critical facilities as recognized by the organization. Performance of this task requires a detailed understanding of an organization's functions, internal infrastructures, and the way these link to external infrastructures.

The purpose of the infrastructure interdependencies assessment is to examine and evaluate the infrastructures (internal and external) that support critical facility functions, along with their associated interdependencies and vulnerabilities.

Risk Characterization

Risk characterization provides a framework for prioritizing recommendations across all task areas. The recommendations for each task area are judged against a set of criteria to help prioritize the recommendations and assist the organization in determining the appropriate course of action. It provides a framework for assessing vulnerabilities, threats, and potential impacts (determined in the other tasks). In addition, the existing risk analysis and management process at the organization should be reviewed and, if appropriate, utilized for prioritizing recommendations. The degree to which corporate risk management includes security factors is also evaluated.

VULNERABILITY ASSESSMENT PROCEDURES

Vulnerability assessment procedures can be conducted using various methodologies. For example, the checklist analysis is an effective technology. In addition, Pareto analysis (80/20 principle), relative ranking, pre-removal risk assessment (PRRA), change analysis, failure mode and effects analysis (FMEA), fault tree analysis, event tree analysis, what-if analysis, and hazard and operability (HAZOP) can be used in conducting the assessment.

Based on our experience, the what-if analysis and HAZOP seem to be the most user-friendly methodologies. A sample what-if analysis procedural outline is presented below, followed by a brief explanation of and outline for conducting HAZOP.

What-If Analysis Procedure/Sample What-If Questions

The steps in a what-if checklist analysis are as follows:

1. Select the team (personnel experienced in the process)
2. Assemble information (piping and instrumentation drawings [P&IDs], process flow diagrams [PFDs], operating procedures, equipment drawings, etc.)
3. Develop a list of what-if questions
4. Assemble your team in a room where each team member can view the information
5. Ask each what-if question in turn and determine
   - What can cause the deviation from design intent that is expressed by the question?
   - What adverse consequences might follow?
   - What are the existing design and procedural safeguards?
   - Are these safeguards adequate?
   - If these safeguards are not adequate, what additional safeguards does the team recommend?
6. As the discussion proceeds, record the answers to these questions in tabular format.
7. Do not restrict yourself to the list of questions that you developed before the project started. The team is free to ask additional questions at any time.

8. When you have finished the what-if questions, examine the checklist. The purpose of this checklist is to ensure that the team has not forgotten anything. While you are reviewing the checklist, other what-if questions may occur to you.

9. Be sure that you follow up all recommendations and action items that arise from the hazards evaluation.

HAZOP Analysis

The HAZOP analysis technique uses a systematic process to (1) identify possible deviations from normal operations and (2) ensure that safeguards are in place to help prevent accidents. The HAZOP uses special adjectives (such as speed, flow, pressure, etc.) combined with process conditions (such as "more," "less," "no," etc.) to systematically consider all credible deviations from normal conditions. The adjectives, called guide words, are a unique feature of HAZOP analysis.

In this approach, each guide word is combined with relevant process parameters and applied at each point (study node, process section, or operating step) in the process that is being examined.

Guide Words	Meaning
No	Negation of the design intent
Less	Quantitative decrease
More	Quantitative increase
Part Of	Other materials present by intent
As Well As	Other materials present unintentionally
Reverse	Logical opposite of the intent
Other Than	Complete substitution

Common HAZOP Analysis Process Parameters

Flow	Time	Frequency	Mixing
Pressure	Composition	Viscosity	Addition
Temperature	pH	Voltage	Separation
Level	Speed	Information	Reaction

The following is an example of creating deviations using guide words and process parameters:

Guide Words		Parameter		Deviation
No	+	Flow	=	No Flow
More	+	Pressure	=	High Pressure
As Well As	+	One Phase	=	Two Phase
Other Than	+	Operation	=	Maintenance
More	+	Level	=	High Level

Guide words are applied to both the more general parameters (e.g., react, mix) and the more specific parameters (e.g., pressure, temperature). With the general parameters, it is not unusual to have more than one deviation from the application of one guide word. For example, "more reaction" could mean either that a reaction takes place at a faster rate or that a greater quantity of product results. On the other hand, some combination of guide words and parameters will yield no sensible deviation (e.g., "as well as" with "pressure").

HAZOP Procedure
1. Select the team.
2. Assemble information (P&IDs, PFDs, operating procedures, equipment drawings, etc.).
3. Assemble your team in a room where each team member can view P&IDs.
4. Divide the system you are reviewing into nodes (you can present the nodes, or the team can choose them as you go along).
5. Apply appropriate deviations to each node. For each deviation, address the following questions:
   - What can cause the deviation from design intent?
   - What adverse consequences might follow?
   - What are the existing design and procedural safeguards?
   - Are these safeguards adequate?
   - If these safeguards are not adequate, what does the team recommend?
6. As the discussion proceeds, record the answers to these questions in tabular format.

NUCLEAR POWER PLANT VULNERABILITY ASSESSMENT PROCESS
The VA for general industry sector entities discussed above can be used as a template—because it is an iterative process due to various interdependencies—for any industry, including nuclear power. Generally, however, the methodology used for nuclear power plant vulnerability assessments is structured as a series of the following seven steps.

1. **Define VA project**—Step 1 provides guidance on the composition of the site team tasked with performing the VA and defining the assessment scope.
2. **Identify critical assets**—Step 2 identifies a master list of nuclear power plant critical assets related to public health and safety.
3. **Screen critical assets for threat susceptibility**—Step 3 leads the VA team to confirm threat characterizations and determine whether any additional threats (based on site-specific factors) should be added to the spectrum.

4. **Define threat-asset scenarios**—Many of the individual threat-asset pairs identified in step 3 can be grouped together to form a single threat scenario.
5. **Characterize the risk of each threat-asset scenario**—Step 5 provides the methodology and guidance for risk characterization of the threat-asset scenarios identified in step 4.
6. **Screening of low-risk scenarios**—Step 6 quantitatively screens low-risk scenarios using criteria typical of that employed in PRA evaluations.
7. **Risk reduction**—Step 7 provides guidance for the evaluation of threat scenarios' potential risk reduction.

The VA methodology provides an approach to evaluate a spectrum of threats, targets, and plant configurations. The VA should be a living assessment that can be updated in the future to account for security and plant modifications, new threats or scenarios, and specific intelligence reports.

REFERENCES AND RECOMMENDED READING

Belke, J. C. 2001. Chemical accident risks in U.S. industry—A preliminary analysis of accident risk data from U.S. hazardous chemical facilities, in *Proceedings of the 10th International Symposium on Loss Prevention and Safety Promotion in the Process Industries*, Stockholm, Sweden. Amsterdam: Elsevier Science.

CBO. 2004. *Homeland security and the private sector*. www.cbo.gov/ftpdocs (accessed May 2, 2008).

Clark, R. M., and R. A. Deininger. 2000. Protecting the nation's critical infrastructure: The vulnerability of U.S. water supply systems, *Journal of Contingency Crisis Management* 8(2): 76–80.

DHS. 2009. National infrastructure protection plan. http://www.dhs.gov/xlibrary/assets/ NIPP_Plan.pdf (accessed May 1, 2009).

DOE/EPRI. 2004. *Probabilistic consequence analysis of security threats—A prototype vulnerability assessment process for nuclear power plants*. Washington, DC: DOE; Palo Alto, CA: EPRI.

FBI. 2009. Terrorism threats to energy sector: Congressional testimony of FBI director. http:// www.fib.gov/congress/congress03/muller021103.htm (accessed May 1, 2009).

FEMA. 2008. *FEMA452 - Risk assessment: A how to guide*. fema.gov/library/file?type= published/filetofile (accessed May 1, 2008).

Foster, S. S. D. 1987. Fundamental concepts in aquifer vulnerability, pollution risk and protection strategy. In *Vulnerability of Soil and Groundwater Pollutants*, ed. W. Van

Duijvenboden & H.G. van Waegeningh. The Hague, The Netherlands: TNO Committee on Hydrological Research Proceedings and Information #38.

GAO. 2004. *RMP-covered industrial processes and off-site consequences of worst-case chemical releases.* Washington, DC: Homeland Security: GAO-04-482T Chemical Security.

GAO. 2005. *Wastewater facilities: Experts' view on how federal funds should be spent.* Washington, DC: U.S. Government Accountability Office, GAO-05-165.

Minter, J. G. 1996. Prevention chemical accidents still a challenge. *Occupational Hazards,* September.

Spellman, F. R. 1997. *A guide to compliance for PSM/RMP.* Lancaster, PA: Technomic.

U.S. DOE. 2002. *Vulnerability assessment methodology: Electric power infrastructure.* Washington, DC: DOE.

U.S. EPA. 2002. *Vulnerability assessment fact sheet.* EPA 816-F-02-025. www.epa.gov/ogwdw/security/index.html (accessed May 2006).

6

Preparation: Security Assessment

We must take the battle to the enemy, disrupt his plans, and confront the worst threats before they emerge.

—*George W. Bush*

Governor Tom Ridge points out the security role for public service professionals: "Americans should find comfort in knowing that millions of their fellow citizens are working every day to ensure our security at every level—federal, state, county, municipal. These are dedicated professionals who are good at what they do. I've seen it up close, as Governor of Pennsylvania. . . . But there may be gaps in the system. The job of the Office of Homeland Security will be to identify those gaps and work to close them.

"Now, obviously, the further removed we get from September 11, I think the natural tendency is to let down our guard. Unfortunately, we cannot do that. The government will continue to do everything we can to find and stop those who seek to harm us. And I believe we owe it to the American people to remind them that they must be vigilant, as well" (Henry 2002).

The physical security of nuclear power plants and their vulnerability to deliberate acts of terrorism was elevated to a national security concern following the 9/11 attacks. Since then, Congress has repeatedly focused oversight and legislative attention on nuclear power plant security requirements established and enforced by the Nuclear Regulatory Commission (NRC) (CRS 2009).

INTRODUCTION*

Because of the seriousness of the threat of terrorism to the nation's nuclear power sector and the enormous economic implications of such attacks, U.S. NRC, U.S. DOE,

---

* Information in this chapter is from *U.S. NRC Frequently Asked Questions about Security Assessments at Nuclear Power Plants* (2009). http://www.nrc.gov/security/gaq-security-access-nuc-pwr-plants.htm (accessed November 22, 2009).

U.S. EPA, and other agencies have worked nonstop since 9/11 in gathering and providing as much advice and guidance as possible to aid nuclear power sector personnel
in protecting nuclear power sector production facilities and associated critical support
infrastructure. In this chapter, we provide an overview of U.S. NRC's response to the
9/11 attacks. The NRC first issued advisories and then orders that required nuclear
power plant licensees to provide specific enhanced capabilities to respond to a terrorist
attack. The NRC then began an accelerated security and engineering review based on
the 9/11 events. The review looked at what could possibly happen if terrorists used an
aircraft to attack a nuclear power plant. Additionally, NRC reviews assessed the potential consequences of other types of terrorist attacks. The NRC analyzed what would
happen as a result of such attacks and what other factors might affect the possibility or
magnitude of a radiation release.

As part of this security review, the NRC conducted detailed engineering studies of a
number of nuclear power plants. These studies assessed the capabilities of these plants
to withstand deliberate attacks involving large commercial aircraft. The NRC studies
included national experts from Department of Energy laboratories who used state-of-
the-art experiments, structural analyses, and fires analyses. The studies at the specific
facilities confirmed that the plants are robust. In addition, the studies found that even
in the unlikely event of a radiological release due to a terrorist attack, there would be
time to implement the required off-site planning strategies already in place to protect
public health and safety.

In a series of three phases, the NRC and the nuclear industry are analyzing the ability of nuclear power plants to withstand damage to or loss of large areas of the plant.
The damage may be caused by a range of deliberate attacks that result in large fires
and explosions.

THREATS AND INCIDENTS

Nuclear power sector threats and incidents may be of particular concern due to the
range of potential consequences:

- Creating an adverse impact on public health within a population
- Disrupting system operations and interrupting the supply of energy
- Causing physical damage to system infrastructure
- Reducing public confidence in the nuclear power supply system
- Long-term denial of nuclear power and the cost of replacement

Keep in mind that some of these consequences would be realized only in the
event of a successful terrorist incident; however, the mere threat of terrorism can
also have an adverse impact on industries that depend on a safe, steady supply of

nuclear power. In addition, the economic implications of such attacks are potentially enormous.

While it is important to consider the range of possibilities associated with a particular threat, assessments are typically based on the probability of a particular occurrence. Determining probability is somewhat subjective and is often based on intelligence and previous incidents.

Threat Warning Signs

To this point we have discussed NRC's recommended and required guidelines. *A threat warning is an occurrence or discovery that indicates a potential threat that triggers an evaluation of the threat.* It is important to note that these warnings must be evaluated in the context of typical industry activity and previous experience in order to avoid false alarms. Following is a brief description of potential warnings.

- **Security breach.** Physical security breaches, such as unsecured doors, open hatches, and unlocked/forced gates, are probably the most common threat warnings. In most cases, the security breach is likely related to lax operations or typical criminal activity such as trespassing, vandalism, and theft. However, it may be prudent to assess any security breach with respect to the possibility of attack.
- **Witness account.** Awareness of an incident may be triggered by a witness account of tampering. Nuclear power sector sites/facilities should be aware that individuals observing suspicious behavior near energy sector facilities/plants will likely call 911 and not the plant. In this case, the incident warning technically might come from law enforcement, as described below. Note: The witness may be a plant employee engaged in normal duties.
- **Direct notification by perpetrator.** A threat may be made directly to the nuclear power sector site, plant, or facility, either verbally or in writing. Historical incidents would indicate that verbal threats made over the phone are more likely than written threats. While the notification may be a hoax, threatening an energy sector unit is a crime and should be taken seriously.
- **Notification by law enforcement.** A nuclear power sector site/facility may receive notification about a threat directly from law enforcement, including local, county, state, or federal agencies. As discussed previously, such a threat could be a result of suspicious activity reported to law enforcement, either by a perpetrator, a witness, or the news media. Other information, gathered through intelligence or informants, could also lead law enforcement to conclude that there may be a threat to the nuclear power sector site/facility. While law enforcement will have to lead in the criminal investigation, the energy sector site/facility has primary responsibility for the safety of its equipment, radiation-producing materials and processes, and public health.

Thus, the plant's role will likely be to help law enforcement to appreciate the public health implications of a particular threat as well as the technical feasibility of carrying out a particular threat.

- **Notification by news media.** A threat to destroy a nuclear power site/facility might be delivered to the news media, or the media may discover a threat. A conscientious reporter would immediately report such a threat to the police, and either the reporter or the police would immediately contact the nuclear power sector site/facility. This level of professionalism would provide an opportunity for the plant to work with the media and law enforcement to assess the credibility of the threat before any broader notification is made.

- **Public health notification.** In this case, the first indication that energy blackouts or energy sector site/facility emergencies (e.g., sabotage) have occurred is the appearance of victims in local emergency rooms and health clinics. Energy sector sites/facilities may therefore be notified, particularly if the cause is unknown or linked to energy materials (oil, gas, coal, or by-products and/or petrochemicals). An incident triggered by a public health notification is unique in that at least a segment of the population has been exposed to a harmful substance. If this agent is a hazardous petrochemical, then the time between exposure and onset of symptoms may be on the order of hours, and thus there is the potential that the contaminant is still present.

OVERVIEW OF REACTOR SECURITY*

Physical security at nuclear power plants involves the threat of radiological sabotage— a deliberate act against a plant that could directly or indirectly endanger public health and safety through exposure to radiation. The Nuclear Regulatory Commission (NRC) establishes security requirements at U.S. commercial nuclear power plants based on its assessment of plant vulnerabilities to, and the consequences of, potential attacks. The stringency of NRC's security requirements and its enforcement program have been a significant congressional issue, especially since the 9/11 terrorist attacks on the U.S.

Nuclear plant security measures are designed to protect three areas of vulnerability: controls on the nuclear chain reaction, cooling systems that prevent hot nuclear fuel from melting even after the chain reaction has stopped, and storage facilities for highly radioactive spent nuclear fuel. U.S. plants are designed and built to prevent dispersal of radioactivity, in the event of an accident, by surrounding the reactor in a steel-reinforced concrete containment structure.

NRC requires commercial nuclear power plants to have a series of physical barriers and a trained security force, under regulations already in place prior to the 9/11 attacks

---

* Material presented in this section is from *Nuclear Power Plant Security and Vulnerabilities* (2009). Congressional Research Service, www.crs.gov.

(10 CFR 73—Physical Protection of Plants and Materials; see chapter 9). The plant sites are divided into three zones: an "owner-controlled" buffer region, a "protected area," and a "vital area." Access to the protected areas is restricted to a portion of plant employees and monitored visitors, with stringent access barriers. The vital area is further restricted, with additional barriers and access requirements. The security force must comply with NRC requirements on prehiring investigations and training (found in 10 CFR 73.55).

A fundamental concept in NRC's physical security requirements is the design basis threat (DBT), which establishes the severity of the potential attacks that a nuclear plant's security force must be capable of repelling. The DBT includes such characteristics as the number of attackers, their training, and the weapons and tactics they could use. Specific details are classified. Critics of nuclear plant security have contended that the DBT should be strengthened to account for potentially large and more sophisticated terrorist attacks.

Reactor vulnerability to deliberate aircraft crashes has also been a major concern since 9/11. Most existing nuclear power plants were not specifically designed to withstand crashes from large jetliners, although analyses differ as to the damage that could result. NRC has determined that commercial aircraft crashes are beyond the DBT but voted in February 2009 to require that new reactor designs be able to withstand such crashes without releasing radioactivity. Nuclear power critics have called for retrofits of existing reactors as well.

Since the 9/11 attacks, NRC and Congress have taken action to increase nuclear power plant security. NRC issued a series of security measures beginning in 2002, including a strengthening of the DBT and establishing the Office of Nuclear Security and Incident Response (NSIR). The office centralizes security oversight of all NRC-regulated facilities, coordinates with law enforcement and intelligence agencies, and handles emergency planning activities. In 2004, NRC implemented a program to conduct "force-on-force" security exercises overseen by NSIR at each nuclear power plant at least every three years. The Energy Policy Act of 2005 (P.L. 109-58) required NRC to further strengthen the DBT, codified the force-on-force program, and established a variety of additional nuclear plant security measures. In December 2008, NRC approved a series of security regulations that require power plants to prepare cyber security plans, develop strategies for dealing with the effects of aircraft crashes, strengthen access controls, improve training for security personnel, and take other new security measures.

DESIGN BASIS THREAT (DBT)

The design basis threat (DBT) profiles the type, composition, and capabilities of adversaries that nuclear plants and nuclear fuel cycle facilities must defend against to prevent radiological sabotage and theft of strategic special nuclear material. NRC licensees use the DBT as the basis for implementing defensive strategies of a specific nuclear plant site through security plans, safeguard contingency plans, and guard training and qualification plans.

General requirements for the DBT are prescribed in NRC regulations (10 CFR § 73.1), while specific attributes of potential attackers, such as their weapons and ammunition, are contained in classified adversary characteristics documents (ACDs).

Fundamental policies on nuclear plant security threats date back to the Cold War. In 1967, the Atomic Energy Commission (AEC) instituted a rule that nuclear plants are not required to protect against an attack directed by an "enemy of the United States"; it was feared that Cuba might launch an attack on Florida reactors (GAO 2006). That so-called enemy of the state rule specifies that nuclear power plants are

> Not required to provide for design features or other measures for the specific purpose of protection against the effects of (a) attacks and destructive acts, including sabotage, directed against the facility by an enemy of the United States, whether a foreign government or other person, or (b) use or deployment of weapons incident to U.S. defense activities. (10 CFR § 50.13)

The Nuclear Regulatory Commission (NRC), the AEC's successor regulatory agency, says that the rule "was primarily intended to make clear that privately-owned nuclear facilities were not responsible for defending against attacks that typically could only be carried out by foreign military organizations" (FR 2007a). NRC's initial DBT, established in the later 1970s, was intended to be consistent with the enemy of the state rule, which remains in effect.

However, the 9/11 attacks drew greater attention to the potential severity of credible terrorist threats. Following the attack, NRC evaluated the extent to which nuclear plant security forces should be able to defend against such threats and ordered a strengthening of the DBT, along with other security measures, on April 29, 2003. That order changed the DBT to "represent the largest reasonable threat against which a regulated private guard force should be expected to defend under existing law," according to the NRC announcement (FR 2003).

In the Energy Policy Act of 2005 (EPACT), Congress imposed a statutory requirement on the NRC to initiate rule making for revising the design basis threat (P.L. 109-58 2005). EPACT required NRC to consider twelve factors in revising the DBT, such as an assessment of various terrorist threats, sizable explosive devices and modern weapons, attacks by persons with sophisticated knowledge of facility operations, and attacks on spent fuel shipments.

NRC approved its final rule amending the DBT (10 CFR Part 73.1) on January 29, 2007, effective April 18, 2007 (FR 2007a). Although specific details of the revised DBT were not released to the public, in general the final rule

- clarifies that physical protection systems are required to protect against diversion and theft of fissile material
- expands the assumed capabilities of adversaries to operate as one or more teams and attack from multiple entry points
- assumes that adversaries are willing to kill or be killed and are knowledgeable about specific target selection
- expands the scope of vehicles that licensees must defend against to include water vehicles and land vehicles beyond four-wheel-drive type
- revises the threat posed by an insider to be more flexible in scope
- adds a new mode of attack from adversaries coordinating a vehicle bomb assault with another external assault

The DBT final rule excluded aircraft attacks, a decision that raised considerable controversy. In approving the rule, NRC rejected a petition from the Union of Concerned Scientists to require that nuclear plants be surrounded by aircraft barriers made of steel beams and cables (the so-called beamhenge concept). Critics of the rule charged that deliberate aircraft crashes were a highly plausible mode of attack, given the events of 9/11. However, NRC contended that power plants were already required to mitigate the effects of aircraft crashes and the "active protection against airborne threats is addressed by other federal organizations, including the military" (NRC 2007a). Additional NRC action on aircraft threats is discussed below.

NRC commissioners in January 2009 rejected a proposal by the NRC staff to strengthen the classified portion of the DBT to include additional capabilities by potential attackers, according to news reports. The staff proposal lost in a 2–2 vote, with one commissioner position currently vacant. In an interview afterward, NRC chairman Dale Klein said the vote could be reconsidered after completion of an ongoing interagency study (Beattie 2009).

Critics of NRC's security regulations have pointed out that licensees are required to employ only a minimum of five security personnel on duty per plant, which they argue is not enough for the job (10 CFR 73.55 [h][3]). Nuclear spokespersons responded that the actual security force for the nation's sixty-five nuclear plant sites numbers more than five thousand, an average of about seventy-five per site (covering multiple shifts). Nuclear plant security forces are also supposed to be aided by local law enforcement officers if an attack occurs.

LARGE AIRCRAFT CRASHES

Nuclear power plants were designed to withstand hurricanes, earthquakes, and other extreme events. But deliberate attacks by large airliners loaded with fuel, such as those that crashed into the World Trade Center and Pentagon, were not analyzed when design requirements of today's reactors were determined (Meserve 2002). Concern about aircraft crashes was intensified by a taped interview shown September 10, 2002, on the Arab TV station al Jazeera, which contained a statement that al Qaeda initially planned to include a nuclear plant in its list of 2001 attack sites.

In light of the possibility than an air attack might penetrate the containment structure of a nuclear plant or a spent fuel storage facility, some interest groups have suggested that such an event could be followed by a meltdown or spent fuel fire and widespread radiation exposure.

Nuclear industry spokespersons have countered by pointing out that relatively small, low-lying nuclear power plants are difficult targets for attack and have argued that penetration of the containment is unlikely and that even if such penetration occurred, it probably would not reach the reactor vessel. They suggest that a sustained fire, such as that which melted the steel support structures in the World Trade Center buildings, would be impossible unless an attacking plane penetrated the containment completely, including its fuel-bearing wings. According to former NRC chair Nils Diaz, NRC studies "confirm that the likelihood of both damaging the reactor core and releasing radioactivity that could affect public health and safety is low" (Diaz 2004).

NRC proposed in October 2007 to amend its regulations to require newly designed power reactors to take into account the potential effects of the impact of a large commercial aircraft (FR 2007b). As discussed in the previous section, NRC considers an aircraft attack to be beyond the design basis threat that plants must be able to withstand, so the requirements of the proposed rule were intended to provide an additional margin of safety. The proposed rule would affect only new reactor designs not previously certified by NRC, because the previous designs were still considered adequately safe. Nevertheless, Westinghouse submitted changes in the certified design of its AP1000 reactor to NRC on May 29, 2007, proposing to line the inside and outside of the reactor's concrete containment structure with steel plate to increase resistance to aircraft penetration (MacLachlan 2007).

Under NRC's 2007 proposed rule, applicants for new certified designs or for new reactor licenses using uncertified designs would have been required to assess the effects that a large aircraft crash would have on the proposed facilities. Each applicant would then describe how the plant's design features, capabilities, and operations would avoid or mitigate the effects of such a crash, particularly on core cooling, containment integrity, and spent fuel storage pools.

In response to comments, the NRC staff proposed in October 2008 that the aircraft impact assessments be conducted by all new reactors, including those using previously certified designs (NRC 2008a). The NRC commissioners, in a 3–1 vote, approved the change February 17, 2009, and added specific design requirements that all new reactors would have to meet (NRC 2009a):

Each applicant subject to this section shall perform a design-specific assessment of the effects on the facility of the impact of a large, commercial aircraft. Using realistic analyses, the applicant shall identify and incorporate into the design those design features and functional capabilities to show that, with reduced use of operative actions: (A) the reactor core remains cooled, or the containment remains intact, and (B) spent fuel cooling or spent fuel pool integrity is maintained.

As noted above, NRC rejected proposals that existing reactors—in addition to new reactors—be required to protect against aircraft crashes, such as by adding beamhenge barriers. However, NRC did impose some additional requirements related to aircraft crashes on all reactors after the 9/11 attacks. In 2002, NRC ordered all nuclear power plants to develop strategies to mitigate the effects of large fires and explosions that could result from aircraft crashes or other causes (NRC 2008a). As part of a broad security rule-making effort, NRC proposed in October 2006 to incorporate the 2002 order on fire and explosion strategies into its security regulations (10 CFR Part 73; NRC 2006). In response to comments, NRC published a supplemental proposed rule in April 2008 to move the fire and explosion requirements into its reactor licensing regulations at 10 CFR Part 50, along with requirements that reactors establish procedures for responding to specific aircraft threat notifications (NRC 2008b). Those regulations received final approval by the NRC commissioners on December 17, 2008 (NRC 2008c).

FORCE-ON-FORCE EXERCISES

EPACT codified an NRC requirement that each nuclear power plant conduct security exercises every three years to test its ability to defend against the design basis threat. In these "force-on-force" exercises, monitored by NRC, a mock adversary force formed outside the plant attempts to penetrate the plant's vital area and simulate damage to key safety components. Participants in the tightly controlled exercises carry weapons modified to fire only blanks and laser bursts to simulate bullets, and they wear laser sensors to indicate hits. Other weapons and explosives, as well as destruction or breaching of physical security barriers, may also be simulated. While one squad of the plant's guard force is participating in a force-on-force exercise, another squad is also on duty to maintain normal plant security. Plant defenders know that a mock attack will take place sometime during a specific period of several hours, but they do not know that the attack scenario will be. Multiple attack scenarios are conducted over several days of exercises.

Full implementation of the force-on-force program began in late 2004. Standard procedures and other requirements have been developed for using the force-on-force exercises to evaluate plant security and as a basis for taking regulatory enforcement action. Many trade-offs are necessary to make the exercises as realistic and consistent as possible without endangering participants or regular plant operations and security.

The NRC required the nuclear industry to develop and train a "composite adversary force" comprising security officers from many plants to simulate terrorist attacks in the force-on-force exercise. However, in September 2004 testimony, GAO criticized the industry's selection of Wackenhut, a security company that guards about half of the U.S. nuclear plants, to also provide the adversary force. In addition to training "questions about the force's independence," GAO noted that Wackenhut had been accused of cheating on previous force-on-force exercises by the Department of Energy (GAO 2004b). Exelon terminated its security contracts with Wackenhut in late 2007 after guards at the Peach Bottom reactor in York County, Pennsylvania, were discovered sleeping while on duty (Executive resigns 2008). EPACT requires NRC to "mitigate any potential conflict of interest that could influence the results of a force-on-force exercise, as the Commission determines to be necessary, and appropriate." NRC's 2007 annual security report to Congress found that the industry adversary team "continued to meet expectations for a credible, well-trained, and consistent mock adversary force" (NRC 2008d).

The first three-year cycle of force-on-force exercises was completed for all U.S. nuclear plant sites at the end of calendar year 2007 (NRC 2008d). During that period, 172 force-on-force exercises were conducted (an average of three per site), and ten security inadequacies were cited. Two of the exercises, both in 2007, resulted in simulated damage or destruction of a vital target by the adversary team. If such an attack had been real, the plant could have released unacceptable levels of radioactivity. Both cases resulted from "the failure of licensee armed security personnel to interpose themselves between the mock adversary and the vital areas and target set components," according to NRC's 2007 security report to Congress. In response to the failures, NRC imposed "immediate compensatory measures followed by long-term corrective actions." Follow-up force-on-force exercises were conducted to verify that the necessary security improvements had been made (NRC 2008d).

SPENT FUEL STORAGE

When no longer capable of sustaining a nuclear reaction, highly radioactive "spent" nuclear fuel is removed from the reactor and stored in a pool of water in the reactor building and at some sites later transferred to dry casks on the plant grounds. Because both types of storage are located outside the reactor containment structure, particular concern has been raised about the vulnerability of spent fuel to attack by aircraft or other means. If terrorists could breach a spent fuel pool's concrete walls and drain the cooling water, the spent fuel's zirconium cladding could overheat and catch fire.

The National Academy of Sciences (NAS) released a report in April 2005 that found that "successful terrorist attacks on spent fuel pools, though difficult, are possible," and that "if an attack leads to a propagating zirconium cladding fire, it could result in the release of large amounts of radioactive material." NAS recommended that the hottest spent fuel be interspersed with cooler spent fuel to reduce the likelihood of fire and that water-spray systems be installed to cool spent fuel if pool water were lost. The report also called for NRC to conduct more analysis of the issue and consider earlier movement of spent fuel from pools into dry storage (NAS 2005). The FY2006 Energy and Water Development Appropriations Act (P.L. 109-103, H. Rept. 109-275) provided twenty-one million dollars for NRC to carry out the site-specific analyses recommended by NAS.

NRC has long contended that the potential effects of terrorist attacks are too speculative to include in environmental studies for proposed spent fuel storage and other nuclear facilities. However, the U.S. Court of Appeals for the Ninth Circuit ruled in June 2006 that terrorist attacks must be included in the environmental study of a dry storage facility at California's Diablo Canyon nuclear plant. NRC reissued the Diablo Canyon study May 29, 2007, to comply with the court ruling, but it did not include terrorism in other recent environmental studies (Beattie 2007).

Long-term management of spent nuclear fuel is currently undergoing review, but spent fuel stored at reactor sites is expected to be moved eventually to central storage, permanent disposal, or reprocessing facilities. Large-scale transportation campaigns would increase public attention to NRC transportation security requirements and related security issues.

SECURITY PERSONNEL AND OTHER ISSUES

After video recordings of inattentive security officers at the Peach Bottom (PA) nuclear power plant were aired on local television, NRC inspection in late September 2007 confirmed that there had been multiple occasions on which multiple security officers were inattentive (NRC 2007b). However, after a follow-up inspection into security issues at the Peach Bottom plant, run by Exelon Nuclear, the NRC concluded that the plant's security program had not been significantly degraded as a result of the guards' inattentiveness. NRC issued a bulletin December 12, 2007, requiring all nuclear power plants to provide written descriptions to their "managerial controls to deter and address inattentiveness and complicity among licensee security personnel" (NRC 2007c).

The incident drew harsh criticism from the House Committee on Energy and Commerce. "The NRC's stunning failure to act on credible allegations of sleeping security guards, coupled with its unwillingness to protect the whistleblower who uncovered the problem, raises troubling questions," said Representative John D. Dingell, then chairman of the committee (CEC 2008). NRC proposed a $65,000 fine on Exelon Nuclear on January 6, 2009 (NRC 2009b).

Following the 9/11 terrorist attacks, NRC conducted a "top-to-bottom" review of its nuclear power plant security requirements. On February 25, 2002, the agency issued "interim compensatory security measures" to deal with the "generalized high-level threat environment" that continued to exist, and on January 7, 2003, it issued regulatory orders that tightened nuclear plant access. On April 29, 2003, NRC issued orders to restrict security officer work hours, establish new security force training and qualification requirements, and increase the DBT that nuclear security forces must be able to defend against, as mentioned previously.

In October 2006, NRC proposed to amend the security regulations and add new security requirements that would codify the series of orders issued after 9/11 and respond to requirements in the Energy Policy Act of 1005 (FR 2006). The new security regulations were approved by the NRC commissioners on December 17, 2008, with the following provisions (Decker 2009):

- **Safety and security interface**—Explicit requirements are established for nuclear plants to ensure that necessary security measures do not compromise plant safety.
- **Mixed-oxide fuel**—Enhanced physical security requirements are established to prevent theft or diversion of plutonium-bearing mixed-oxide (MOX) fuel.
- **Cyber security**—Nuclear plants must submit security plans to prevent cyber attacks on digital computer and communications systems and networks. The cyber security plan will become a license condition for each plant.
- **Aircraft attack mitigative strategies and response**—Nuclear plants must prepare strategies for responding to warnings of an aircraft attack and for mitigating the effects of large explosions and fires.
- **Plant access authorization**—Nuclear plants must implement more rigorous programs for authorizing access, including enhanced psychological assessments and behavioral observation.
- **Security personnel training and qualification**—Modifications to security personnel requirements include additional physical fitness standards, increased minimum qualification scores in mandatory personnel tests, and requirements for on-the-job training.
- **Physical security enhancements**—New requirements are intended to ensure availability of backup security command centers and uninterruptible power supplies to detection systems, enhanced video capability, and protection from waterborne vehicles.

A proposal by NRC staff to release more details about the results of nuclear plant security inspections was defeated by the NRC commissioners in a 2–2 vote on January 21, 2009. Under current policy, NRC announces after a security inspection whether any violations that were found were of low safety significance or moderate or higher safety significance. Critics of the current policy contend that the public needs more detail to be assured of plant security. The policy's supporters counter that greater information about security inspection findings could inadvertently provide useful information to terrorists (Weil 2009).

REFERENCES AND RECOMMENDED READING

AWWARF. 2002. *Online monitoring for drinking water utilities.* AWWA Research Foundation and CRS PORAQUA. Denver, CO: American Water Works Association.

Beattie, J. 2007. NRC takes two roads on terror. *Energy Daily*, review issue.

———. 2009. NRC chairman questions case for tougher DBT. *Energy Daily*, 1.

CEC. 2008. *Energy and Commerce Committee to probe breakdown in NRC oversight* (January 7), http://energycommerce.house. gov/Press_110 /11Onr 149.shtml.

CRS. 2009. *Nuclear power plant security and vulnerabilities.* Congressional Research Service, www.crs.gov.

Decker, D. 2009. CRS personal communication, NRC Office of Congressional Affairs, February 27.

Diaz, N. 2004. Letter from Nils Diaz to Secretary of Homeland Security Tom Ridge.

Executive resigns in storm over sleeping guards. 2008. *Washington Post*, January 10.

Fenichell, S. 1996. *Plastic: The making of a synthetic century.* New York: Harper Business.

FR. 2003. All operating power reactor licensees; order modifying licenses. *Federal Register* 68(88).

———. 2006. NRC, power reactor security requirements. *Federal Register* 71(207).

———. 2007a. Design basis threat. *Federal Register* 72.

———. 2007b. Consideration of aircraft impacts for new nuclear power plant reactor designs. *Federal Register* 72(171).

GAO. 2004a. *Homeland security—chemical security.* Washington, DC: GAO-04-482T.

———. 2004b. Nuclear Regulatory Commission: Preliminary observations on efforts to improve security at nuclear power plants. Statement of Jim Wells, Director, Natural Resources and Environment to the Subcommittee on National Security, Emerging Threat, and International Relations, House Committee on Government Reform, 14.

———. 2006. *Nuclear power plants—efforts made to upgrade security, but the Nuclear Regulatory Commission's design basis threat process should be improved.* GAO-06-388. Washington, DC: GAO.

Henry, K. 2002. New face of security. *Government Security*, 30–37.

MacLachlan, A. 2007. Westinghouse changes AP1000 design to improve plane crash resistance. *Nucleonics Week* (June 21).

Meserve, R. A. 2002. NRC chairman, research: Strengthening the foundation of the nuclear industry. Speech to Nuclear Safety Research Conference.

NAS. 2005. Board on radioactive waste management, safety and security of commercial spent nuclear fuel storage. Public report released April 6 by National Academy of Sciences.

NRC 2006. Power reactor security requirements, proposed rule. *Federal Register* 71.

———. 2007a. NRC approves final rule amending security requirements. *News Release* No. 07-012.

———. 2007b. NRC commences follow-up safety inspection at Peach Bottom (November 5). http://www.nrc.gov/reading-rm/doc-collections/news/2007/07-057.i.html.

———. 2007c. *Security officer attentiveness.* NRC Bulletin 2007-1, December 12.

———. 2008a. *Final rule—Consideration of aircraft impacts for new nuclear power reactors.* Rulemaking Issue Affirmation, SECY-08-0152.

———. 2008b. Power reactor security requirements, supplemental proposed rule. *Federal Register* 73.

———. 2008c. NRC approves final rule expanding security requirements for nuclear power plants. Press release available at http://www.nrc.gov/reading-rm/doc-collections/new/2008-227.html.

———. 2008d. Office of Nuclear Security and Incident Response, Report to Congress on the security inspection program for commercial power reactor and category 1 fuel cycle facilities: Results and status update: Annual report for calendar year 2007. NUREG-1885 (July), 7, http://www.nrc.gov/reading-rm/doc-collections/congress-docs/correspondence/2008/boxer-07-01-2008.pdf.

———. 2009a. *Final rule—Consideration of aircraft impacts for new nuclear power reactors.* Commission Voting Record, SECY-08-0152.

———. 2009b. *NRC News Releases.* http://www.nrc.gov/reading-rm/doc-collections/news/2009 (accessed October 22, 2010).

———. 2009c. *Frequently asked questions about security assessments at nuclear power plants.* http://www.nrc.gov/security/faq-security-access-nuc-pwr-plants.html (accessed November 22, 2009).

PL. 109-58 2005. *Title VI, Subtitle D—Nuclear Security* (Secs. 651-657). Sec. 651 adds Atomic Energy Act Sec. 170E, design basis threat rule making.

U.S. EPA. 2001. *Protecting the nation's water supplies from terrorist attack: Frequently asked questions.* Washington, DC: United States Environmental Protection Agency.

———. 2003. *Response protocol toolbox: Planning for and responding to drinking water contamination threats and incidents.* Washington, DC: United States Environmental Protection Agency.

Weil, J. 2009. Commissioners reach statement on security-related amendment. *Inside NRC,* February 2.

WHO. 2003. *Public health response to biological and chemical weapons: WHO guidance.* 2nd ed.

7

Cyber Security

Society is growing increasingly dependent upon large-scale, highly distributed systems that operate in unbounded network environments. Unbounded networks, such as the Internet, have no central administrative control and no unified security policy. Furthermore, the number and nature of the nodes connected to such networks cannot be fully known. Despite the best efforts of security practitioners, no amount of system hardening can assure that a system that is connected to an unbounded network will be invulnerable to attack. The discipline of survivability can help ensure that such systems can deliver essential services and maintain essential properties such as integrity, confidentiality, and performance, despite the presence of intrusion.

—*R. J. Ellison et al., 1997*

In order for proper corrective action to be effected, two things must occur: fear and then outrage.

—*F. R. Spellman, 2010*

CYBERTERRORISM: WHAT IS IT?*

Thirteen months after the 9/11 attacks, in a college classroom where a 400/500-level risk management course was being taught, students were asked, without warning, the following question:

Define cyberterrorism.

A few of the students' responses include:

"That's when all the lights go out."

"That's when all hell breaks loose."

---

* Much of the information in this chapter is based on information in F. R. Spellman and N. E. Whiting, 2004. *Safety Engineering: Principles and Practices*, 2nd ed. Lanham, MD: Government Institutes.

"That's when the trains wreck."

"No more Twitter . . . no more reason to live!"

"Well, there goes my Madden football. . . ."

"That's the all-time suck-a-roo, man."

"That's when the planes don't fly."

"No TV . . . no phone . . . no Facebook . . . no nothing!"

"No more texting, sexting, or whatever!"

"Well, that'll cure my Blackberry thumb."

"That's when we all freeze to death in the dark."

"Nobody has ever been killed by a cyberterrorist . . . so who cares . . . just give me a Stephen King novel and I am happy . . . gee, give me two of 'em and I'm real happy!"

"That's digital Pearl Harbor Day."

"That's digital Pearl Harbor Day?" Well, we will come back to this statement, but for now we need a more precise definition of cyberterrorism. In the first place, it is important to point out that according to a CRS (2003a) report to Congress, no single definition of the term *terrorism* has gained universal acceptance. Additionally, no single definition of the term *cyberterrorism* has been universally accepted. Also, labeling a computer attack as cyberterrorism is problematic because it is often difficult to determine the intent, identity, or political motivations of a computer attacker with any certainty (was the attack mischievous or malicious?) until long after the event has occurred.

The Department of Homeland Security (DHS) defines cyberterrorism as "a criminal act perpetrated through computers resulting in violence, death and/or destruction, and creating terror for the purpose of coercing a government to change its policies" (Dick 2002).

Others have defined cyberterrorism as the politically motivated use of computers as weapons or as targets, by subnational groups or clandestine agents intent on violence, to influence an audience or cause a government to change its policies. The definition may be extended by noting that DOD operations for informative warfare (CRS 2003b) also include physical attacks on computer facilities and transmission lines.

With regard to the term *digital Pearl Harbor*, this has been used since 1991. Berinato (2003) points out that the term has seen common usage. He also points out that *Newsweek*, in an article in 1998, suggested it would come in the form of a "sophisticated attack on our digital workings [which] could create widespread misery: everything from power failures to train wrecks." Today, the term is rather stale from overuse; even the former cyber security czar Richard Clarke declared that "digital Pearl Harbors are happening every day" (Berinato 2003).

We can readily distinguish between a mischievous digital happening, such as the implanting of a computer virus, and a full-scale, totally unpredictable and malicious digital Pearl Harbor attack simply by weighing the impact of one or the other. Actually, Berinato (2003) has already done that. Consider the following, for example:

BERINATO'S DIGITAL PEARL HARBOR (A MOMENT THAT WILL LIVE IN CYBER INFAMY)

Date: December 7, Any Year

Five factors distinguish the digital Pearl Harbor from the virus attacks we've suffered to date.

1. First, event disrupts all backups systems.
2. Second, event leads to cascading failures.
3. Third, though the attack is instantaneous, its aftereffects linger for weeks.
4. Fourth, after it is over, the attack's origin is pinpointed and the vulnerability it exploited is determined.
5. Last, once the source of the event is determined, it's revealed that the loss of property and life was completely, absolutely, and tragically avoidable.

PLANNING A COMPUTER ATTACK

According to CRS (2003a), there are five basic steps traditionally used by computer hackers to gain unauthorized access and subsequently take over computer systems. These five steps may be used to plan a computer attack for purposes of cyber crime or cyber espionage and may also be employed for purposes of cyberterror. The steps are frequently automated through use of special hacker tools that are freely available to anyone via the Internet. Using these basic steps, often supplemented with automated intrusion tools, attackers have successfully taken over computer sites and remained undetected for long periods of time (Skoudis 2002). Highly skilled hackers use automated tools that are also highly sophisticated, and their effects are initially much more difficult for computer security staff and technology to detect. These sophisticated hacker tools are usually shared only among an exclusive group of other highly skilled hacker associates.

Step 1. Reconnaissance

In this first step, hackers employ extensive preoperative surveillance to find out detailed information about an organization that will help them later gain unauthorized access to computer systems. The most common method is social engineering, or tricking an employee into revealing sensitive information (such as a telephone number or a password). Other methods include dumpster diving, or rifling through an organization's trash to find sensitive information (such as floppy disks or important documents that have not been shredded). This step can be automated if the attacker installs on an office computer a virus, worm, or "spyware" program that performs surveillance and then transmits useful information, such as passwords, back to the attacker. "Spyware" is a form of malicious code that is quietly installed on a computer without user knowledge when a user visits a malicious website. It may remain undetected by firewalls or current anti-virus security products while monitoring keystrokes to record web activity or collect snapshots of screen displays and other restricted information for transmission back to an unknown third party.

Step 2. Scanning

Once in possession of special restricted information, or a few critical phone numbers, an attacker performs additional surveillance by scanning an organization's computer software and network configuration to find possible entry points. This process goes slowly, sometimes lasting months, as the attacker looks for several vulnerable openings into a system.

Step 3. Gaining Access

Once the attacker has developed an inventory of software and configuration vulnerabilities on a target network, he or she may quietly take over a system and network by using a stolen password to create a phony account or by exploiting a vulnerability that allows the attacker to install a malicious Trojan horse, or automatic "bot" that will await further commands sent through the Internet.

Step 4. Maintaining Access

Once attackers have gained unauthorized access, they may secretly install extra malicious programs that allow them to return as often as they wish. These programs, known as rootkits or backdoors, run unnoticed and can allow attackers to secretly access a network at will. If an attacker can gain all the special privileges of a system administrator, then the computer or network has been completely taken over and is "owned" by the attacker. Sometimes the attacker will reconfigure a computer system or install software patches to close the previous security vulnerabilities just to keep other hackers out.

Step 5. Covering Tracks

Sophisticated attackers desire quiet, unimpeded access to the computer systems and data they take over. They must stay hidden to maintain control and gather more intelligence or to refine preparations to maximize damage. The rootkit or Trojan horse programs often allow the attacker to modify the log files of the computer system or to create hidden files to help avoid detection by the legitimate system administrator. Security systems may not detect the unauthorized activities of a careful intruder for a long period of time.

CORRECTIVE ACTIONS

The safety and security of a nuclear power plant's computer system has elements that demand the full attention of the person in charge of safety and security. Familiarity with important issues in computer system and data security is critical. While safety and security personnel shouldn't be expected to know as much about computers as a systems engineer, ensuring that the IT (information technology) staff has the tools it needs to provide reasonable levels of system security, ensuring that security measures are in place to limit access to data and equipment, securing SCADA (see chapter 8), and keeping up with current issues in computer and data security is a growing part of the safety and security professional's job. Nuclear facility administration, IT, and the safety and security engineer should work together on formulating plant policy on computer access, security, auditing, and backup systems. Current best practice for information security is a layered, defense-in-depth program that involves policy, procedures, and tools.

Best Practices: Layered Security for Defense-in-Depth

According to Larry McCreary of IT Security, as networks become increasingly complex, security has become much more than off-the-shelf software. Today firewalls, authentication, intrusion detection/prevention, and web filtering are all part of an effective data security program. Best practices include:

- A regularly updated and reviewed security policy as a formal condition of employment
- A general security approach that allows access only to authenticated users
- Encryption of sensitive data accessible by public connections
- Well-configured and enforced access rules across physical, logical, and social boundaries
- Traffic control across physical, logical, and social boundaries
- Role-based authentication/authorization (as needed), including dynamic passwords (generated by hardware or software tokens or sent to mobile devices, etc.) for network administrators, remote users, and employees accessing mission-critical information.

- Filtering content that could cause:
  - Network vulnerabilities (e-mail attachments, hacker tools, viruses, etc.)
  - Legal liability (pornography, criminal skills, anonymizers, cults, etc.)
  - Poor network availability (MP3, streaming content, etc.)
  - Regular third-party audits and assessments

NRC PHYSICAL SECURITY REGULATIONS (10 CFR PART 73)

As you might expect, the Nuclear Regulatory Commission (NRC 2009) requires each licensee currently licensed to operate nuclear power facilities to develop, submit for approval, and put into effect a cyber security plan that satisfies the requirements specified in its Physical Security Regulations, outlined in 10 CFR part 73.54 (protection of digital computer and communication systems and networks).

NRC (2009) 73.54 states that (a) each licensee subject to the requirements of this section shall provide high assurance that digital computer and communication systems and networks are adequately protected against cyber attacks. Specifically

(1) The licensee shall protect digital computer and communication systems and networks associated with:
   (i)   Safety-related and important-to-safety functions
   (ii) Security functions
   (iii) Emergency preparedness functions, including offsite communications
(2) The licensee shall protect the systems and networks from cyber attacks that would:
   (i)   Adversely impact the integrity or confidentiality of data and/or software
   (ii) Deny access to systems, services, and/or data
   (iii) Adversely impact the operation of systems, networks, and associated requirements
(b) To accomplish this, the licensee shall:
   (1) Analyze digital computer and communication systems and networks and identify those assets that must be protected against cyber attacks
   (2) Establish, implement, and maintain a cyber security program for the protection of the cyber assets and incorporate the cyber security program as a component of the physical protection program
(c) The cyber security program must be designed to:
   (1) Implement security controls to protect cyber assets from cyber attacks
   (2) Apply and maintain defense-in-depth protective strategies to ensure the capability to detect, respond to, and recover from cyber attacks
   (3) Mitigate the adverse effects of cyber attacks
   (4) Ensure that the functions of protected assets are not adversely impacted due to cyber attacks

(d) As part of the cyber security program, the licensee shall:
  (1) Ensure that appropriate facility personnel, including contractors, are aware of cyber security requirements and receive the training necessary to perform their assigned duties and responsibilities
  (2) Evaluate and manage cyber risks
  (3) Ensure that modifications to cyber assets are evaluated before implementation to ensure that the cyber security performance objectives are maintained
(e) The licensee shall establish, implement, and maintain a cyber security plan that implements the cyber security program requirements of this section.
  (1) The cyber security plan must describe how the requirements of this section will be implemented and must account for the site-specific conditions that affect implementation.
  (2) The cyber security plan must include measures for incident response and recovery for cyber attacks. The cyber security plant must describe how the licensee will:
    (i) Maintain the capability for timely detection and response to cyber attacks
    (ii) Mitigate the consequences of cyber attacks
    (iii) Correct exploited vulnerabilities
    (iv) Restore affected systems, networks, and/or equipment affected by cyber attacks
(f) The licensee shall develop and maintain written policies and implementing procedures to implement the cyber security plan. Polices, implementing procedures, site-specific analysis, and other supporting technical information used by the licensee need not be submitted for commission review and approval as part of the cyber security plan but are subject to inspection by NRC staff on a periodic basis.
(g) The licensee shall review the cyber security program as a component of the physical security program in accordance with the requirements of § 73.55(m), including the periodicity requirements.
(h) The licensee shall retain all records and supporting technical documentation required to satisfy the requirements of this section as a record until the commission terminates the license for which the records were developed, and shall maintain superseded portions of these records for at least three (3) years after the record is superseded, unless otherwise specified by the commission.

Cyber/Computer Security Terminology

In the real world of data systems security, a gap exists between theoretical information security best practices and the reality of implementation. In part, under many circumstances, this gap is widened by the IT/digital staff's inability to communicate the necessary information in a way that nontechnical users—from worker level to top administration—can understand. Of course, like any other profession, the digital/ computer industry has a huge and ever-growing body of jargon, and the ability to discuss computer security issues with IT staff relies on understanding these terms; the following section provides a basic vocabulary related to digital/computer security. Moreover, to comply with NRC cyber security regulations and to fashion a commission-approved plan, it is important to first understand terminology related to cyber security and, second, understand the potential threats. For more cyber/computer-oriented terminology, visit www.sans.org/resources/glossary.php.

Definitions

access control—Ensures that resources are granted only to those users who are entitled to them.

access management—Maintenance of access information that consists of four tasks: account administration, maintenance, monitoring, and revocation.

access—The process of interacting with a system. (verb or noun) Users access a system (v). A user gains computer access (n).

account harvesting—The process of collecting all the legitimate account names on a system.

active content—Program code embedded in the contents of a web page that, when accessed by a web browser, automatically downloads and executes on the user's workstation.

activity monitors—System monitor designed to prevent virus infection by monitoring for malicious activity on a system and blocking that activity when possible.

applet—Java programs; an application program that uses the client's web browser to provide a user interface.

audit trail—In computer security systems, a chronological record of when users log in, how long they are engaged in various activities, what they were doing, and whether any actual or attempted security violations occurred. An audit trail may be on paper or on disk.

audit—The independent collection of records to assess their veracity and completeness.

authentication—The process of confirming the correctness of the claimed identity.

backdoor—An entry point to a program or a system that is hidden or disguised, often created by the software's author for maintenance. A certain sequence of control characters permits access to the system manager account. If the backdoor becomes known, unauthorized users (or malicious software) can gain entry and cause damage. Also a tool installed after a compromise to give an attacker easier access to the compromised system around any security mechanisms that are in place.

bandwidth—Commonly used to mean the capacity of a communication channel to pass data through the channel in a given amount of time. Usually expressed in bits per second.

biometrics—Access system that uses physical characteristics of the users to determine access.

bit—The smallest unit of information storage; a contraction of the term "binary digit"; one of two symbols—"0" (zero) and "1" (one)—that are used to represent binary numbers.

broadcast—To simultaneously send the same message to multiple recipients; one host to all hosts on a network.

browser—A client computer program that can retrieve and display information from servers on the World Wide Web.

brute force—A cryptanalysis technique or other kind of attack method involving an exhaustive procedure that tries all possibilities, one by one.

buffer overflow—Occurs when a program or process tries to store more data in a buffer (temporary data storage area) than it was intended to hold. Since buffers are created to contain a finite amount of data, the extra information—which has to go somewhere—may overflow into adjacent buffers, corrupting or overwriting the valid data held in them.

byte—A fundamental unit of computer storage; the smallest addressable unit in a computer's architecture. Usually holds one character of information and usually means eight bits.

cache—Pronounced "cash," a special high-speed storage mechanism. It can be either a reserved section of main memory or an independent high-speed storage device. Two types of caching are commonly used in personal computers: memory caching and disk caching.

ciphertext—The encrypted form of the message being sent.

competitive intelligence—Espionage using legal, or at least not obviously illegal, means.

computer emergency response team (CERT)—An organization that studies computer and network INFOSEC to provide incident response services to victims of attacks, publish alerts concerning vulnerabilities and threats, and offer other information to help improve computer and network security.

computer network—A collection of host computers together with the subnetwork or internetwork through which they can exchange data.

computer security audit—An independent evaluation of the controls employed to ensure appropriate protection of an organization's information assets.

computer security—Technological and managerial procedures applied to computer systems to ensure the availability, integrity, and confidentiality of information managed by the computer system.

cookie—Data exchanged between an HTTP server and a browser (a client of the server) to store information on the client side and retrieve it later for server use. An HTTP server, when sending data to a client, may send along a cookie, which the client retains after the HTTP connection closes. A server can use this mechanism to maintain persistent client-side state information for HTTP-based applications, retrieving the state information in later connections.

corruption—A threat action that undesirably alters system operation by adversely modifying system functions or data.

cost-benefit analysis—A cost-benefit analysis compares the cost of implementing countermeasures with the value of the reduced risk.

cryptanalysis—The mathematical science that deals with analysis of a cryptographic system to gain knowledge needed to break or circumvent the ciphertext to plaintext without knowing the key.

cryptography—Cryptography garbles a message in such a way that anyone who intercepts the message cannot understand it.

data aggregation—The ability to get a more complete picture of the information by analyzing several different types of records at once.

data custodian—The entity currently using or manipulating the data and, therefore, temporarily taking responsibility for the data.

data-driven attack—A form of attack in which the attack is encoded in innocuous-seeming data that is executed by a user or other software to implement an attack. In the case of firewalls, a data-driven attack is a concern since it may get through the firewall in data form and launch an attack against a system behind the firewall.

data encryption standard (DES)—A widely used method of data encryption using a private (secret) key. Seventy-two quadrillion (72,000,000,000,000,000) or more possible encryption keys can be used. For each given message, the key is chosen at random from among this enormous number of keys. Like other private key cryptographic methods, both the sender and the receiver must know and use the same private key.

decryption—The process of transforming an encrypted message into its original plaintext.

defense in-depth—The approach of using multiple layers of security to guard against failure of a single security component to ensure that each system on the network is secured to the greatest possible degree. May be used in conjunction with firewalls.

denial of service (DoS)—The prevention of authorized access to a system resource or the delaying of system operations and functions.

dictionary attack—An attack that tries all of the phrases or words in a dictionary in trying to crack a password or key. A dictionary attack uses a predefined list of words compared with a brute force attack, which tries all possible combinations.

disassembly—The process of taking a binary program and deriving the source code from it.

disaster recovery plan (DRP)—A program and system for the recovery of IT systems in the event of a disruption or disaster.

domain hijacking—An attack by which an attacker takes over a domain by first blocking access to the domain's DNS server and then putting his own server up in its place.

domain name—A name that locates an organization or other entity on the Internet.

domain—A sphere of knowledge, or a collection of facts about some program entities, or a number of network points or addresses, identified by a name. On the Internet, a domain consists of a set of network addresses. In the Internet's domain name system, a domain is a name with which server records are associated that describe subdomains or hosts. In Windows NT and Windows 2000, a domain is a set of network resources (applications, printers, and so forth) for a group of users. The user need only log into the domain to gain access to the resources, which may be located on a number of different servers in the network.

due care—A minimal level of protection in place in accordance with industry best practice.

due diligence—The requirement that organizations must develop and deploy a protection plan to prevent fraud and abuse and deploy a means to detect them if they occur.

dumpster diving—Obtaining passwords and corporate directories by searching through discarded media.

egress filtering—Filtering outbound traffic.

e-mail bombs—Code that when executed sends many messages to the same address(es) for the purpose of using up disk space and/or overloading the e-mail or web server.

encryption—The process of scrambling files or programs, changing one character string to another through an algorithm (such as the DES algorithm).

escrow passwords—Passwords recorded and stored in a secure location (like a safe) and used by emergency personnel when privileged personnel are unavailable.

event—An observable occurrence in a system or network.

exposure—A threat action whereby sensitive data is directly released to an unauthorized entity.

false reject—An authentication system's failure to recognize a valid user.

fault line attacks—Attacks that use weaknesses between interfaces of systems to exploit gaps in coverage.

fault tolerance—A design method that ensures continued systems operation in the event of individual failures by providing redundant system elements.

file transfer protocol (FTP)—A TCP/IP protocol specifying the transfer of text or binary files across the network.

filtering router—An internetwork router that selectively prevents the passage of data packets according to a security policy, often used as a firewall or part of a firewall. A router usually receives a packet from a network and decides where to forward it on a second network. A filtering router does the same, but it first decides whether the packet should be forwarded at all, according to some security policy. The policy is implemented by rules (packet filters) loaded into the router.

fingerprinting—Sending strange packets to a system to gauge how it responds in order to determine the operating system.

firewall—A system or combination of systems that enforces a boundary between two or more networks.

flooding—An attack that attempts to cause a failure in (especially, in the security of) a computer system or other data processing entity by providing more input than the entity can process properly.

fork bomb—Attack that uses the fork() call to create a new process that is a copy of the original. By doing this repeatedly, all available processes on the machine can be taken up.

gateway—A network point that acts as an entrance to another network.

generic utilities—General-purpose code and devices, including screen grabbers and sniffers that look at data and capture information such as passwords, keys, and secrets.

global security—The ability of an access control package to permit protection across a variety of mainframe environments, providing users with a common security interface to all.

hackers—Those intent upon entering an environment to which they are not entitled entry for whatever purpose (entertainment, profit, theft, prank, etc.). Hackers usually use iterative techniques, escalating to more advanced methodologies and use of devices to intercept the communications property of another.

handheld—A small electronic device used to collect data.

hardening—The process of identifying and fixing vulnerabilities in a system.

hijack attack—A form of active wiretapping in which the attacker seizes control of a previously established communication association.

honey pot—Programs that simulate one or more network services that you designate on your computer's ports. An attacker assumes you're running vulnerable services that can be used to break into the machine. A honey pot can be used to log access attempts to those ports, including the attacker's keystrokes. This could give you advance warning of a more concerted attack.

host—Any computer that has full two-way access to other computers on the Internet, or a computer with a web server that serves the pages for one or more web sites.

host-based security—The technique of securing an individual system from attack. Host-based security is operating system and version dependent.

hot standby—A backup system configured in such a way that it may be used if the system goes down.

hybrid attack—Attack that builds on the dictionary attack method by adding numerals and symbols to dictionary words.

hyperlink—In hypertext or hypermedia, an information object (such as a word, a phrase, or an image, usually highlighted by color or underscoring) that points (indicates how to connect) to related information located elsewhere that can retrieved by activating the link.

incident handling—An action plan for dealing with intrusions, cyber theft, denial of service, fire, floods, and other security-related events, comprising a six-step process—preparation, identification, containment, eradication, recovery, and lessons learned.

incident—An adverse network event in an information system or network or the threat of the occurrence of such an event.

incremental backups—Backup system that duplicates only files modified since the previous backup.

ingress filtering—Filtering inbound traffic.

input validation attacks—Intentionally sending unusual input in the hopes of confusing an application.

insider attack—An attack originating from inside a protected network.

integrity—The need to ensure that information has not been changed accidentally or deliberately and that it is accurate and complete.

Internet protocol (IP)—The method or protocol by which data is sent from one computer to another on the Internet.

Internet standard—A specification, approved by the IESG and published as an RFC, that is stable and well understood, is technically competent, has multiple, independent, and interoperable implementations with substantial operational experience, enjoys significant public support, and is recognizably useful in some or all parts of the Internet.

interrupt—A signal that informs the operating system that something has occurred.

intranet—A computer network, especially one based on Internet technology, that an organization uses for its own internal and usually private purposes and that is closed to outsiders.

intrusion detection—A security management system that gathers and analyzes information from various areas within a computer or a network to identify possible security breaches, including both intrusions (attacks from outside the organization) and misuse (attacks from within the organization).

IP flood—A denial of service attack that sends a host more echo request ("ping") packets than the protocol implementation can handle.

IP sniffing—Stealing network addresses by reading the packets. Harmful data is then sent stamped with internal trusted addresses.

IP splicing—An attack whereby an active, established session is intercepted and co-opted by the attacker. IP splicing attacks may occur after an authentication has been made, permitting the attacker to assume the role of an already authorized user.

IP spoofing—The technique of supplying a false IP address to illicitly impersonate another system by using its IP network address.

ISSA—Information Systems Security Association.

issue-specific policy—Policy to address specific needs within an organization, such as a password policy.

list-based access control—Associates a list of users and their privileges with each object.

log clipping—The selective removal of log entries from a system log to hide a compromise.

log retention—How long audit logs are retained and maintained.

logging—The process of storing information about events that occurred on the firewall or network.

loopback address—Pseudo IP addresses that always refer back to the local host and are never sent out onto a network.

malicious code—Software (e.g., Trojan horse) that appears to perform a useful or desirable function but actually gains unauthorized access to system resources or tricks a user into executing other malicious logic.

malware—A generic term for a number of different types of malicious code.

masquerade attack—A type of attack in which one system entity illegitimately poses as (assumes the identity of) another entity.

monoculture—A large number of users who run the same software and are thus vulnerable to the same attacks.

National Institute of Standards and Technology (NIST)—A unit of the U.S. Commerce Department that promotes and maintains measurement standards. It has active programs for encouraging and assisting industry and science to develop and use these standards.

natural disaster—Any "act of God" (e.g., fire, flood, earthquake, lightning, or wind) that disables a system component.

network mapping—To compile an electronic inventory of the systems and the services on a network.

network taps—Hardware devices that hook directly onto the network cable and send a copy of the traffic that passes through it to one or more other networked devices.

network worm—A program or command file that uses a computer network as a means for adversely affecting a system's integrity, reliability, or availability, A network worm may attack from one system to another by establishing a network connection. It is usually a self-contained program that does not need to attach itself to a host file to infiltrate network after network.

network-level firewall—A firewall in which traffic is examined at the network protocol packet level.

null session—Known as Anonymous Logon, it is a way of letting an anonymous user retrieve information such as user names and shares over the network or connect without authentication.

octet—A sequence of eight bits. An octet is an eight-bit byte.

operating system—System software that controls a computer and its peripherals. Modern operating systems such as Windows 95 and NT handle many of a computer's basic functions.

Orange Book—The Department of Defense Trusted Computer System Evaluation Criteria. It provides information to classify computer systems, defining the degree of trust that may be placed in them.

overload—Hindrance of system operation by placing excess burden on the performance capabilities of a system component.

partitions—Major divisions of the total physical hard disk space.

password—A secret code assigned to a user and known by the computer system. Knowledge of the password associated with the user ID is considered proof of authorization.

password authentication protocol (PAP)—A simple, weak authentication mechanism where a user enters the password and it is then sent across the network, usually in the clear.

password cracking—The process of attempting to guess passwords, given the password file information.

password sniffing—Passive wiretapping, usually on a local area network, to gain knowledge of passwords.

patch—A small update released by a software manufacturer to fix bugs in existing programs.

penetration—Gaining unauthorized logical access to sensitive data by circumventing a system's protections.

penetration testing—Used to test the external perimeter security of a network or facility.

perimeter-based security—The technique of securing a network by controlling access to all entry and exit points of the network.

personal firewalls—Firewalls installed and run on individual PCs.

PIN—In computer security, a personal identification number used during the authentication process. Known only to the user.

plaintext—Ordinary, readable text before being encrypted into ciphertext or after being decrypted.

policy—Organization-level rules governing acceptable use of computing resources, security practices, and operational procedures.

polymorphism—The process by which malicious software changes its underlying code to avoid detection.

port scan—A series of messages sent to learn which computer network services, each associated with a "well-known" port number, the computer provides. This computer cracker favorite gives the assailant an idea where to probe for weaknesses. By sending a message to each port, one at a time, the response received indicates whether the port is used and can therefore be probed for weakness.

program policy—A high-level policy that sets the overall tone of an organization's security approach.

promiscuous mode—When a machine reads all packets off the network, regardless of who they are addressed to. This mode is used by network administrators to diagnose network problems but also by unsavory characters to eavesdrop on network traffic (which might contain passwords or other information).

proprietary information—Information unique to a company and its ability to compete, such as customer lists, technical data, product costs, and trade secrets.

protocol—A formal specification for communicating. Protocols exist at several levels in a telecommunication connection.

proxy server—A server that acts as an intermediary between a workstation user and the Internet so that the enterprise can ensure security, administrative control, and caching service. A proxy server is associated with or part of a gateway server that separates the enterprise network from the outside network and a firewall server that protects the enterprise network from outside intrusion.

race condition—Exploits the small window of time between when a security control is applied and when the service is used.

reconnaissance—The phase of an attack where attackers find new systems, map out networks, and probe for specific, exploitable vulnerabilities.

registry—In Windows operating systems, the central set of settings and information required to run the Windows computer.

remote access—The hookup of a remote computing device via communications lines such as ordinary phone lines or wide area networks to access network applications and information.

resource exhaustion—Attacks that tie up finite resources on a system, making them unavailable to others.

reverse engineering—Acquiring sensitive data by disassembling and analyzing the design of a system component.

reverse lookup—Finding out the hostname that corresponds to a particular IP address. Reverse lookup uses an IP (Internet Protocol) address to find a domain name.

risk analysis or assessment—The analysis of an organization's information resources, existing controls, and computer system vulnerabilities. It establishes a potential level of damage in dollars and/or other assets.

risk—The product of the level of threat with the level of vulnerability. It establishes the likelihood of a successful attack.

rogue program—Any program intended to damage programs or data.

rootkit—A collection of tools (programs) that a hacker uses to mask intrusion and obtain administrator-level access to a computer or computer network.

router—Device to interconnect logical networks by forwarding information to other networks based upon IP addresses.

scavenging—Searching through data residue in a system to gain unauthorized knowledge of sensitive data.

secure shell (SSH)—A program that lets its user log into another computer over a network, execute commands in a remote machine, and move files from one machine to another.

security policy—A set of rules and practices that specify or regulate how a system or organization provides security services to protect sensitive and critical system resources.

sensitive information—As defined by the federal government, any unclassified information that, if compromised, could adversely affect the national interest or conduct of federal initiatives.

server—The control computer on a local area network that controls software access to workstations, printers and other parts of the network.

shadow password file—A system file in which encryption user passwords are stored so that they aren't available to people who try to break into the system.

share—A resource made public on a machine, such as a directory (file share) or printer (printer share).

smurf—An attack that works by spoofing the target address and sending a ping to the broadcast address for a remote network, which results in a large number of ping replies being sent to the target.

sniffer—A tool that monitors network traffic as it is received in a network interface.

sniffing—A synonym for "passive wiretapping."

social engineering—A euphemism for nontechnical or low-technology means, including lies, impersonation, tricks, bribes, blackmail, and threats, used to attack information systems or gain illicit access to systems. Typically carried out by telephoning users or operators pretending to be authorized users.

spam—Electronic junk mail or junk newsgroup postings.

spoof—Attempt by an unauthorized entity to gain access to a system by posing as an authorized user.

stack mashing—Using a buffer overflow to trick a computer into executing arbitrary code.

stealthing—Approaches used by malicious code to conceal its presence in the infected system.

system security officer (SSO)—A person responsible for enforcement or administration of the security policy that applies to the system.

system-specific policy—A policy written for a specific system or device.

tamper—To deliberately alter a system's logic, data, or control information to cause the system to perform unauthorized functions or services.

TCO—Total cost of ownership, a model to help IT professionals understand and manage the direct and indirect costs of acquiring, maintaining, and using an application or a computing system, including training, upgrades, and administration as well as the purchase price.

TCP/IP—A synonym for "Internet Protocol Suite;" of which the Transmission Control Protocol and the Internet Protocol are important parts. The basic communication language or protocol of the Internet, TCP/IP is often used as a communications protocol in a private network (either an intranet or an extranet).

threat—A potential for violation of security that exists when a circumstance, capability, action, or event could breach security and cause harm.

threat assessment—The identification of types of threats that an organization might be exposed to.

threat model—Used to describe a given threat and the harm it could to do a system if the system has a vulnerability.

threat vector—The method a threat uses to get to the target.

transmission control protocol (TCP)—A set of rules (protocol) used along with the Internet Protocol to send data in the form of message units between computers over the Internet. While IP takes care of handling the actual delivery of the data, TCP takes care of keeping track of the individual units of data (called packets) that a message is divided into for efficient routing through the Internet. While the IP protocol deals only with packets, TCP enables two hosts to establish a connection and exchange streams of data. TCP guarantees delivery of data and also guarantees that packets will be delivered in the same order in which they were sent.

Trojan horse—A computer program that appears to have a useful function but that also carries a hidden and potentially malicious function for invading security mechanisms, sometimes by exploiting legitimate authorizations of a system entity that invokes the program.

turn commands—Commands inserted to forward mail to another address for interception.

uniform resource locator (URL)—The global address of documents and other resources on the World Wide Web.

unprotected share—In Windows terminology, a share is a mechanism that allows a user to connect to file systems and printers on other systems. An unprotected share is one that allows anyone to connect to it.

user contingency plan—The alternative methods of continuing business operations if IT systems are unavailable.

user—A person, organization entity, or automated process that accesses a system, whether authorized to do so or not.

virus—A hidden, self-replicating section of computer software, usually malicious logic that propagates by infecting—that is, inserting a copy of itself into and becoming part of—another program. A virus cannot run by itself; it requires that its host program be run to make the virus active.

vulnerability—A flaw or weakness in a system's design, implementation, or operation and management that could be exploited to violate the system's security policy.

war dialing—A simple means of trying to identify modems susceptible to compromise in attempting to circumvent perimeter security.

web server—A software process that runs on a host computer connected to the Internet, intended to respond to HTTP requests for documents from client web browsers.

wiretapping—Monitoring and recording data flowing between two points in a communication system.

worm—A computer program that can run independently, propagating a complete working version of itself onto other hosts on a network and consuming computer resources destructively.

wrap—Using cryptography to provide data confidentiality service for a data object.

Threats to Cyber/Computer System Security

Risks to cyber/computer systems data include physical theft, logical usurpation, application damage through viral attack, and "social engineering" attacks. For example, the possibility is very high that any organization with more than twenty workers will have lost laptops or handhelds in the past year. The associated costs for such losses are usually determined only for replacement costs—which doesn't include the value of the lost data—both the monetary and security value of the data. Count up all of the replacement costs, including equipment, software, and restoring or replacing data, and the reasoning behind why staff should be taught clear policies, procedures, and the importance of cyber/computer security and the presence of risks becomes clear.

Computer system risks mean that any complete security program pays attention to all risk areas. That means developing ways to protect

- The physical equipment itself, including desktop computers (whether office based or home computers supplied by the company) and the associated peripherals (output devices like printers, input devices like scanners, and data transfer, backup, and storage devices, like Zip drives or CD or DVD drives, whether external or internal); laptop computers and their associated gear; and an ever-widening body of handheld devices, including PDAs, bar-code readers, and many different types of data readers.
- Logical assets, including IP and other network configuration information
- Application and data integrity
- Password integrity

Different types of equipment present different security needs. Laptops are easier to physically steal than desktop units but are more apt to be the targets of incidental theft, not a deliberate plan to steal a particular unit because of the data it contains. More important is the protection of the computer system and the information held within the system, which may have value or may simply be vulnerable to hackers.

Developing the Security Process

When developing a computer and data protection program, vulnerability management is a primary concern. A vulnerability management process should be put in place to ensure that all desktops remain secure from potential vulnerabilities. According to Trinity Security Services (2003), the various stages of the process include

- **Inventory**—You can't determine vulnerability of your equipment and systems without knowing what you have. Begin, regularly update, and maintain lists of all equipment, type, manufacturer, capacity, special features, operating system, important applications, and company purpose, operator and other users, with serial numbers. This data can be critically important in tracking missing or stolen equipment, in helping to pinpoint certain problems, and for insurance and replacement purposes.
- **Gather Intelligence**—You can't protect something you don't know is at risk. Being aware of the latest vulnerabilities is key to maintaining a high security posture. Since new dangers can occur at any time, track potential problems specific to your inventory.
- **Audit**—Use automated vulnerability-scanning tools to regularly check system status for known vulnerabilities (This can give you 90 percent protection.) Develop internal auditing practices to increase your protection percentage.
- **Fix**—Once vulnerabilities have been identified, fix them quickly (before they allow damage), through patches, upgrading, removal of software. Test any changes before general release to personnel.
- **Review**—Review the findings and fixes for the process to identify where risk reduction changes are possible. Develop a plan for regular implementation of changes.

WATERFALL LIFE CYCLE SECURITY MODEL

For protecting digital computers in safety systems of nuclear power generating stations, a method must be put in place that will satisfy the NRC's regulations with respect to high functional reliability and design requirements for computers used in safety systems of nuclear power plants.

NRC's *Regulatory Guide 1.152* describes a method that the staff of the NRC (2006) deems acceptable for complying with the commission's regulations for protecting high functional reliability, design quality, and cyber security for the use of digital computers in safety systems of nuclear power plants. In this context, the term *computer* identifies a system that includes computer hardware, software, firmware, and interfaces.

The regulatory position uses the waterfall lifestyle phases (Winston W. Royce, 1929–1996, first termed and described the waterfall model in an article in 1970) as a framework for describing specific digital safety system security guidance. The waterfall model is a sequential software development process in which progress is seen as flowing steadily downward (like a waterfall) through the phases dealing with (for our purposes) concepts, requirements, design, implementation, testing, installation-checkout-acceptance testing, operation, maintenance, and retirement (see figure 7.1). The digital safety system development process should address potential security vulnerabilities in each phase of the digital safety system lifecycle.

The lifecycle phase-specific security requirements should be commensurate with the risk and magnitude of the harm resulting from unauthorized and inappropriate access, use, disclosure, disruption, or destruction of the digital safety system. Regulatory positions, top to bottom, shown in figure 7.1 describe digital safety system security guidance for the individual phases of the lifecycle.

Concepts Phase

In the concepts phase, the licensee and developer should identify safety system security capabilities that should be implemented. The licensee and developer should perform security assessment to identify potential vulnerabilities in the relevant phase of the system life cycle. The results of the analysis should be used to establish security requirements of the system (hardware and software). Remote access to the safety system should not be implemented. Computer-based safety systems may transfer data to other systems through one-way communication pathways.

Requirements Phase

System Features—The licensees and developers should define the security functional performance requirements and system configuration; interfaces external to the system; and the requirements for qualification, human factors engineering, data definitions, documentation for the software and hardware, installation and

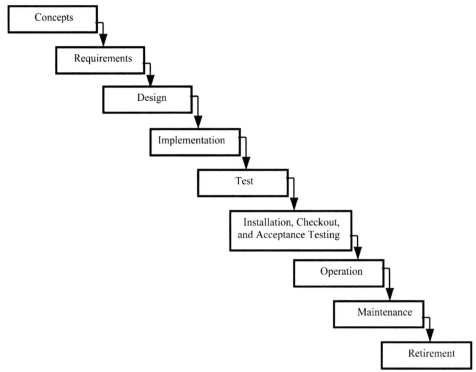

FIGURE 7.1
Phases of NRC's Cyber Security Waterfall Model

acceptance, operation and execution, and maintenance. The security requirements should be part of the overall system requirements. Therefore, the verification and validation (V&V) process of the overall system should ensure the correctness, completeness, accuracy, testability, and consistency of the system security requirements. Requirements specifying the use of predeveloped software and systems (e.g., reuse software and commercial off-the-shelf systems) should address the vulnerability of the safety system (e.g., by using predeveloped software functions that have been tested and are supported by operating experience).

Development Activities—The development process should ensure the system does not contain undocumented code (e.g., backdoor coding), malicious code (e.g., intrusions, viruses, worms, Trojan horses, or bomb codes), and other unwanted and undocumented functions or applications.

Design Phase

System Features—The safety system security requirements identified in the system requirement specification should be translated into specific design configuration items in the system design description. The safety system security design configuration items should address control over (1) physical and logical access to the system functions, (2) use of safety system services, and (3) data communication with other systems. Design configuration items incorporating predeveloped software into the safety system should address security vulnerabilities of the safety system.

Physical and logical access control should be based on the results of cyber security qualitative risk analyses. Cyber security risk is the combination of the consequence to the nuclear power plant and the susceptibility of a digital system to internal and external cyber attack. The results of the analyses may require more complex access control, such as a combination of knowledge (e.g., password), property (e.g., key, smart card), or personal features (e.g., fingerprints), rather than just a password.

Development Activities—The developer should delineate the standards and procedures that will conform with the applicable security policies to ensure the system design products (hardware and software) do not contain undocumented code (e.g., backdoor coding), malicious code (e.g., intrusions, viruses, worms, Trojan horses, or bomb codes), and other unwanted or undocumented functions or applications.

Implementation Phase

In the system (integrated hardware and software) implementation phase, the system design is transformed into code, database structures, and related machine executable representations. The implementation activity addresses hardware configuration and setup; software coding and testing; and communication configuration and setup (including the incorporation of reused software and commercial off-the-shelf (COTS) products).

System Features—The developer should ensure that the security design configuration item transformations from the system design specification are correct, accurate, and compete.

Development Activities—The developer should implement security procedures and standards to minimize and mitigate tampering with the developed system. The developer's standards and procedures should include testing with scanning as appropriate, to address undocumented colds or malicious functions that might (1) allow unauthorized access or use of the system or (2) cause systems to behave beyond the system requirements. The developer should account for hidden functions and vulnerable features embedded in the code, and their purpose and impact on the safety system. If possible, these functions should be disabled, removed, or (as a minimum) addressed (e.g., as part of the failure modes and effects analysis of the application code) to prevent any unauthorized access.

Scanning is dependent on the platform and code being used and may not be available for the specified code and compiler. This may be a difficult task with little assurance that the results will be comprehensive and successful in uncovering hidden problems, given the size and complexity of most modern computer systems. Pure application code scanning may be partially successful, but many operating systems, machine code, and callable library function aspects of the system may not be able to be successfully scanned and are just as likely to be where avenues for exploitation exist.

Commercial off-the-shelf (COTS) computing architecture is likely to be proprietary and generally unavailable for review. It is likely that there is no reliable method to determine security vulnerabilities for operating systems (for example, Microsoft and other operating system suppliers do not provide access to the source code for operating systems and callable code libraries). In such cases, unless such systems are modified by the application developer, the security effort should be limited to ensuring that the features within the system do not compromise the security requirements of the system, and the security functions should not be compromised by the other system functions.

Test Phase

The objective of testing security functions is to ensure that the system security requirements are validated by execution of integration, system, and acceptance tests where practical and necessary. Testing includes system hardware configuration (including all external connectivity), software integration testing, software qualification testing, system integration testing, system qualification testing, and system factory acceptance testing.

System Features—The security requirements and configuration items are part of validation of the overall system requirements and design configuration items. Therefore, security design configuration items are only one element of the overall system validation. Each system security feature should be validated to verify that the implemented system does not increase the risk of security vulnerabilities and does not reduce the reliability of safety functions.

Development Activities—The developer should configure and enable the designed security features correctly. The developer should also test the systems hardware architecture, external communication devices, and configuration for unauthorized pathways and system integrity. Attention should be focused on built-in original equipment manufacturer (OEM) features.

Installation, Checkout, and Acceptance Testing

In installation and checkout, the safety system is installed and tested in the target environment. The system licensee should perform an acceptance review and test the safety system security features. The object of the installation and checkout security testing is to verify and validate the correctness of the safety, physical, and logical system security features in the target environment.

System Features—The licensee should ensure that the system features enable the licensee to perform postinstallation testing of the system to verify and validate that the security requirements have been incorporated into the system appropriately.

Development Activities—A licensee should have a digital system security program. The security policies, standards, and procedures should ensure that installation of the digital system will not compromise the security of the digital system, other systems, or the plant. This may require the licensee to perform a security assessment, which includes a risk assessment, to identify the potential security vulnerabilities caused by installation of the digital system. The risk assessment should include an evaluation of new security constraints in the system; an assessment of the proposed system changes and their impact on system security; and an evaluation of operating procedures for correctness and usability. The results of this assessment should provide a technical basis for establishing certain security levels for the systems and the plant.

Operation Phase

The operation life-cycle process involves the use of the safety system by the licensee in its intended operational environment. During the operations phase, the licensee should ensure that the system security is intact by techniques such as periodic testing and monitoring, review of system logs, and real-time monitoring where possible.

The licensee should evaluate the impact of safety system changes in the operating environment on safety system security; assess the effect on safety system security of any proposed changes; evaluate operating procedures for compliance with the intended use; and analyze security risks affecting the licensee and the system. The licensee should evaluate new security constraints in the system; access proposed system changes and their impact on system security; and evaluate operating procedures for correctness and usability.

Maintenance Phase

The maintenance phase is activated when the licensee changes the system or associated documentation. These changes may be categorized as follows:

- Modifications (i.e., corrective, adaptive, or perfective changes)
- Migration (i.e., the movement of a system to a new operational environment)

- Replacement (i.e., the withdrawal of active support by the operation and maintenance organization, partial or total replacement by a new system, or installation of an upgraded system)

System modifications may be derived from requirements specified to correct errors (corrective), to adapt to a changed operating environment (adaptive), or to respond to additional licensee requests or enhancements (perfective).

Maintenance Activities—Modifications of the safety system should be treated as development processes and should be verified and validated as described above. Security functions should be assessed as described in the above regulatory positions and should be revised (as appropriate) to reflect requirements derived from the maintenance process.

When migrating systems, the licensee should verify that the migrated systems meet the safety system security requirements. The maintenance process should continue to conform to existing safety system security requirements unless those requirements are to be changed as part of the maintenance activity.

Quality Assurance—The licensee should address security in its quality assurance program. The security quality assurance section can be incorporated into the existing quality assurance program. The cyber security features should be maintained under a configuration management program.

The licensee's quality assurance group (such as information/network security expert) should conduct periodic audits to determine the effectiveness of the digital safety system security procedures.

If the safety system security functions were not previously verified and validated using a level of effort commensurate with the safety system security functional requirements, and the appropriate documentation is unavailable or inadequate, the licensee should determine whether the missing or incomplete documentation should be generated. In making this determination of whether to generate missing documentation, the minimum safety system security functional requirements should be taken into consideration.

Incident Response—The licensee should develop an incident response and recovery plan for responding to digital system security incidents (e.g., intrusion, viruses, worms, Trojan horses, or bomb codes). The plan should be developed to address various loss scenarios and undesirable operations of plant digital systems, including possible interruptions in service due to the loss of system resources, data, facility, staff, and/or infrastructure. The plan should define contingencies for ensuring minimal disruption to critical services in these instances.

Audits and Assessments—The licensee should perform periodic computer system security self-assessments and audits, which are key components of a good security program. The licensee should assess proposed safety system changes and their impact on safety system security; evaluate anomalies that are discovered during operation; assess migration requirements; and assess modifications made, including V&V tasks, to ensure that vulnerabilities have not been introduced into the plant environment from modifications.

Retirement Phase

In the retirement life-cycle phase, the licensee should assess the effect of replacing or removing the existing safety system security functions from the operating environment. The licensee should include in the scope of this assessment the effect on safety and nonsafety system interfaces of removing the system security functions. The licensee should document the methodology by which a change in the safety system security functions will be mitigated (e.g., replacement of the security functions, isolation from other safety systems and licensee interactions, or retirement of the safety system interfacing functions). The security procedures should include cleansing the hardware and data. Upon removal from service, the licensee should consider data cleansing, disk destruction, or complete overwrite.

NETWORKS AND STANDALONES

Networked computer systems offer so many benefits that most places are heading in that direction, if they aren't there already. These days, servers and an IT staff handle computer operations at many facilities, and while many facilities might have stand-alone computers dedicated to specific purposes, facilities with only stand-alone computers are rapidly becoming a thing of the past. The differences between stand-alone computers and a networked system make some aspects of data protection easier and others harder.

Server systems allow IT staff to automatically and regularly back up all data, keep track of employee Internet and e-mail usage (in part to protect from data theft or loss) and to help diagnose problems. Servers also allow IT staff to monitor many types of employee computer activity. How far management goes in keeping tabs on worker computer use is up to individual facility policy, but networked systems provide the possibility. Setting up the capability for and gathering that information is possible on stand-alone computers, too, but is much more labor intensive. Duplication of equipment is also an expensive factor for standalones.

One real advantage of standalones is that they are isolated if a computer virus strikes. Even if a viral infection occurs (from a modem or data transfer device), it is isolated and can be confined. Data exchange and backup are much more time consuming, however.

With a stand-alone system, choosing each workstation's peripherals in light of the other stand-alone workstations keeps data transfer and backup both possible and simple.

PORTABLE COMPUTING

Laptops and handheld computing devices offer convenience, portability, flexibility, and many other benefits. They also take both equipment and data out into the cold, cruel world, where they are vulnerable to loss and theft. Personnel who have had equipment lifted from under their noses or who have absentmindedly walked off and left a laptop in a taxi should be more than embarrassed at the loss—because company policy and training should alert them to the possibility of such losses. The bottom line is that whatever the security issue, the biggest threat comes from the user. The best protection is to implement effective and well-managed safe work practices for all employees—education and training is key to protecting the network elements that go into the public domain.

Laptops and handhelds, including PDAs, hold information and provide access to e-mail and other network resources. Even a casual thief (not only one who has deliberately targeted a particular company as a source for valuable information) could benefit from information stored in a poorly secured device. Enforcing standardization and policy among PDA users is especially difficult, since the users are generally the owners of the device. Since they don't see the PDA as a computer and since the PDA is personal, not company, property, workers are sometimes much more careless with them than they would be with a corporate laptop.

Laptop Security

In some ways (especially physically), laptops are more difficult than full-sized desktop computers or workstations to protect. *A Guide to Rolling Out Handhelds in the Corporate Environment* (2009) lists the principle laptop security issues:

- Physical theft
- Loss of information
- Hacking
- Installation of noncorporate software

Physical theft or loss is the most likely security risk for laptops, and the key to preventing such losses is user education. When workers take laptops off-site, either occasionally or as their usual routine, they must be aware that cars, airport lounges, taxis, trains, and hotel rooms are high-risk areas. An unattended laptop is a theft waiting to happen.

Cable locks (similar in many ways to bicycle locks) are used to attach laptops to stationary objects to prevent theft. Newer machines have the capability of logical security measures to protect data from casual thievery—company and user passwords, for example. The data the laptop contains could easily be more valuable (to the facility, if to no one else) than the cost of the equipment, so regular backup before taking any laptop or other portable device off-site should be standard practice. Data encryption provides a deeper level of protection for sensitive data.

Common practice is that laptops are equipped with Internet connection capability as a transport mechanism for connecting to the office. However, using the Internet itself can be risky. A desktop-level firewall protects the laptop during connection to any "foreign" network (i.e., the Internet).

Handheld Devices

Handheld devices offer low cost, ease of use, and portable convenience. While contemporary laptops can have all the power and hard disk capacity of an office PC, a handheld device is a palmtop whose weight is measured in ounces, not pounds. Limited to one or a few dedicated purposes, the rechargeable batteries last three to four times longer than a PC's. Dedicated handhelds (the UPS driver's delivery pad, for example, or a bar-code reader) are less attractive to casual thieves than laptops, too. Handheld devices are a highly effective addition to mobile working practices, especially for data collection. Administrators and sales staff (and others) love their PDAs for the huge amount of useful data such devices can make portable.

The combination of low cost, portability, and high functionality makes handheld computing equipment more and more attractive for companies with the need for a mobile workforce. However, many IT departments haven't fully incorporated handhelds into computer and data security programs. A handheld computing component that addresses personnel, technical, and security issues should be included for facilities that issue such equipment.

Handheld equipment's small size doesn't necessarily mean low capacity. The iPod (now used mostly for music but capable of a startling array of functions for other purposes) can hold thirty gigabytes of memory in a case about the size of but much thinner than a pack of cigarettes. The large data storage capacity of both laptops and handhelds means that staff members have the potential of carrying large amounts of company information around with them. Best practice is to upload data to the server on a regular basis (determined by value of the data and opportunity), not storing it for long periods of time, so that any loss is mostly equipment loss, not data loss. Collecting the data again can be impossible (for time-sensitive measurements, for example) or difficult (would that UPS driver have to go back and collect all those signatures again?). The involved costs are hard to accurately calculate.

Other than loss or theft, the biggest problem areas for handhelds are in file transfer and synchronization. Tools that enable all devices to synchronize to a single server means that information available to mobile users can be easily monitored. Password protection on many devices is two-level—a password for the device and one for individual applications or files is common.

So far, handheld devices are not heavily affected by computer viruses. Currently, only a few Trojans affect them, most of them irritating but harmless—like randomly switching backlights on and off or posting nasty and hard-to-eradicate screen messages. However, sooner or later, creating more troublesome handheld-directed trouble will occur to hackers, so keeping an eye out for information on new vulnerabilities is important. Many hardware vendors have established partnerships with leading anti-virus organizations already.

Table 7.1. Tips for Information Security

Four Fundamentals of Security

1. Identification and authentication	Who are you, and can you prove it ?
2. Access control	Who can access, edit, create, delete, or copy data?
3. Confidentiality and integrity	Who sent this, who can read it, and has the data been changed?
4. Administration and audit	Who did what, when, and where?

Know Your 'Enemies'

1. The malicious insider
2. The thief
3. The hacker
4. Ignorance and complacency

Know What Your Enemies Use

1. Social engineering
2. Password cracking
3. Network monitoring (sniffing)
4. War dialing (brute force modem attacks)

Top Ten Elements of Security

1. Policy	Define, implement, and measure systems, networks, and applications for compliance
2. Education	Part of (1) above is to ensure your users understand the consequences of their actions and the importance of information security to the organization
3. Warnings	Implement online warnings informing users (internal and external) of the rules of access to the systems
4. Assess	Determine what you are trying to protect, why, and what value it has
5. Protect	Based on (4) above, implement firewalls, anti-virus, and intrusion-detection (IDS) tools
6. Audit	Regularly audit systems for access violations, latest applicable patches, and integrity of operating systems and applications
7. Password	Enforce strong password policy and run password-guessing tests to determine those with easy-to-guess passwords
8. Scan	Periodically check for known vulnerabilities and the latest exploits
9. Backup	Ensure regular, clean backups are made to enable speedy recovery
10. Response	Set up an incident response team with complementary measures

Source: Malcolm Skinner of AXENT Technologies (2002). *Tips for Information Security*.. www.axent.com.

Equipment can be infected by malicious code through e-mail attachments or during synchronization with infected systems. Anti-virus software at all synchronization points on the network cuts down the possibility of mobile workers infecting the corporate system and vice versa. In fact, some operating systems can be made resistant to virus attacks, though this can restrict the functionality of the device—in general, the more you increase the security, the more you restrict what the user can do. (Some of the previous information came from *A Guide to Rolling Out Handhelds in the Corporate Environment*, 2009).

THE HUMAN-ERROR FACTOR

People cause the biggest risks to information systems, whether deliberate or accidental. In fact, people—the users themselves—are a commonly overlooked possible breach in any security infrastructure. They'll write down passwords and leave them in obvious places. They'll set their software to remember passwords. They'll open e-mail with strange attachment file endings and cute taglines, and the resulting avalanche of virally induced spamming will bring down the whole system and help to spread the infection at top speed throughout the world. Educating them on the implications of their actions is ongoing, an integral part of employee training. IT staff and safety personnel should work together to educate administrators and the workforce in computer and data protection. Training and worker education helps to control a wide range of problems. For example, as safety engineer, you should know that the workers in your organization have been trained to follow IT instructions on updating viral protection regularly, that they know better than to open suspicious e-mail, that they are aware that a random stranger on-site could be seeking vulnerable equipment or data, and that passwords must never be shared with anyone outside a verified IT staff member. In short, one of the best protections for computer security is a user who is thinking.

Security Measures for IT Staff

While the IT staff handles the background of computer security, they are often the ones responsible for communicating security needs to workers. According to Dancho Danchev (2003), IT security issues handled on a worker level may include

- Physically secure terminal and working places. Make staff aware of the dangers of malicious "snoopers" walking around workplaces or having access to terminals.
- Set terminals for automatic logout, or set up a time-out so that the system is protected once it detects the operator is not in front of the keyboard.
- Limit the use of notes and papers for any sensitive information such as passwords, IPs, and anything that might help a potential intruder gain access to systems. Advise

staff to shred and destroy password notes each time they leave their workplace. This is a well-known workplace weakness, so limit or completely eliminate the use of these notes.

- Remember that the general staff may or may not have real experience with computer use. What seems simple to an IT staff member may be only halfway comprehensible to the person at the workstation—or that person may be totally lost. Effective communication both ways is key to performance.

Internal IT Security Measures

- Conduct a complete security audit of the system before physically connecting it to the network.
- Make sure the system has the latest versions of the software installed and securely configured.
- Consider blocking access to the test system from the Internet during network auditing and testing.
- Verify that the system to connect doesn't contain any sensitive data yet.
- Install an intrusion detection system to see how often the new system is probed for various vulnerabilities.
- Subscribe to appropriate security-related newsletters and mailing lists for updates on the latest vulnerabilities.
- Visit the appropriate websites as an early warning system for potential intrusions from outdated or unpatched software.
- Read the latest security-related white papers as an essential step in self-education (such as Danchev 2003).

System Administrator Mistakes

System administrators manage the continued operation of system computers and network functionality. In many organizations they also handle equipment security, potential and real intrusion detection, and network security. Holding many responsibilities increases workloads, as well as levels of stress and distraction from multiple projects, and can increase the probability of accidental human error. Common system administration problems include

- The lack of a well-established personnel security policy
- Connecting misconfigured systems to the Internet
- Relying on tools alone, without auditing and analysis
- Failing to monitor the logs
- Running extra and unnecessary services/scripts (Danchev 2003)

Plant Executive Mistakes

As top management sees what the Internet offers, they realize that many possibilities are at hand—the potential for new markets, higher profits, and so forth. However, often they don't see that the global connectivity isn't all golden. Connectivity represents a threat to sensitive information without solid system security. Ensuring that executives know the cost of doing business on the Internet includes effective security measures is critical for data security program success. Common mistakes include

- Employing untrained and inexperienced experts
- Failing to realize the consequences of a potential security breach
- Trying to handle the information security issue on the cheap
- Relying too heavily on commercial tools and products
- Thinking that security is a one-time investment (Danchev 2003)

End User Mistakes

While end users can't be blamed for IT or management problems, on both network systems and stand-alone computers, trouble often arrives via a run-of-the-mill user's workstation. Training and communication from both safety personnel and IT staff is essential to keeping systems secure. Workers cause problems by

- Violating the company's security policy
- Forwarding sensitive data to home computers
- Writing down account data
- Downloading contaminated e-mail or from infected websites
- Failing to pay serious attention to physical security (Danchev 2003)

COMPUTER SECURITY AUDITING

Issues and actions for computer and data systems security include many elements to consider. Using a computer security audit checklist provides steps to securing computer and information systems. Just like any other safety area, computer and data security is not static—no once-and-done solution is possible. The equipment to protect, the risks, and the threats change rapidly. This section addresses common audit items and the audit questions to help determine system needs.

- Who within the organization should have access to what information? Ensure that people do not unnecessarily have full access to the systems.
- What information is most important to protect in terms of confidentiality?

- Does the acceptable use policy explain what everyone in the organization should and should not do with the information systems?
- Is a current equipment and configuration list maintained for the organization?
- Is a current software installation list available for all computers and equipment in the organization?
- Are all passwords used in the IT systems changed from the default?
- Are log-on passwords effectively maintained as confidential, changed regularly, and created in a format that cannot easily be guessed?
- Do any workers take laptops and other portable devices off the network and outside of standard security procedures?
- Will any remote workers need to connect to the network?
- What is the organization plan for security requirement evaluation?
- Is someone responsible for network and information security? Does that person have the authority to take action to respond to security breaches and vulnerabilities?
- Can abnormal network activity on your network be identified?
- How is company physical waste disposed of? Secure computer data revealed through print-out paper waste instead of system intrusion is just as disruptive (Rose 2002).

DATA PROTECTION

Once a system is in place, daily maintenance must occur to prevent trouble. Individual tasks must be scheduled to fit the individual needs of the organization and should include a number of different measures and concerns. These include backing up data, ensuring continued operation during high-level problems, and viral protection in a number of ways.

Backup Protection

Both equipment and data are important issues in computer security. While computer equipment is valuable, attractive to thieves, expensive to purchase or replace, and certainly vulnerable, if your system is properly protected by backup, the equipment itself is generally fairly easily replaceable. Often equipment theft is covered by insurance, and the equipment can be replaced. More important than the equipment itself is the data the computer equipment is used to generate, control, store, format, and distribute. The computer equipment is there for a concrete purpose—computers are elaborate tools for data management—and the data is not so easy to replace if irretrievably lost. The downtime related to replacement of both equipment and data is another important area of concern.

Data protection starts with backup. Set up equipment and systems for easy data backup (either automatically, or easy for individual users to perform), and keep duplicate backups off-site as well as on-site. Servers allow automated backup—but automated backup doesn't matter if the place burns down and you don't have a current data backup off-site. Stand-alone computers should be outfitted with a company-standardized means of backing up data, and the backup media should be collected regularly and stored off-site. Obviously, a scrubbable media is more cost effective than many one-time use media (writeable CDs, for example), but currently, prices for writeable and even rewriteable CDs are low. Service agencies on the Internet offer automated backup of data onto remotely located servers over broadband Internet connections—these have advantages and disadvantages, but they are certainly off-site and simple. Be sure you secure your backups as part of your data protection policy.

If facility data isn't adequately backed up, restoration may be difficult, time consuming, and costly. How much of the data in your facility computer system was entered by hand? With a massive loss of data, people may have to go back in and tediously re-enter it, which not only is time consuming but introduces errors as well, making the loss that much more expensive. Was the data read from an electronic reader of some sort, a handheld data collector, or an in-the-field laptop? How much data does that electronic reader or handheld hold, anyway? Loss of either equipment or the data on the equipment is tough to handle.

Backup solutions should meet the requirements of any user, from a periodic (weekly or monthly) copying of some information onto a CD or other external media to a fully replicated server system with two parallel systems operated in separate parts of the world. The more extensive the system is, the more apt it is to be properly backed up, but the bigger the tangle of data can be to sort out. Essential for any backup system are regular tests of the retrievability of the backed up data. IT should keep updated lists of all the programs on each workstation's hard drive. Data files without the application used to generate or read the files are more difficult to use or useless. Don't rely on individual workers to know the ins and outs of their workstation's applications. IT should maintain records for each workstation and all the related devices of the system. The backup storage system should also ensure that the stored data is traceable back to the workstation it was copied from.

Factors to consider in arranging backups include

- What information or data should be backed up?
- How often should I back up information?
- What is the most efficient and cost-effective backup medium for my system?
- Can I restore information in the event of a loss of the computer(s), the backup machine, and the backup software?
- Can I restore data when I check to make sure my backup processes are working?
- Where will I store my backups? Are the backups themselves secure?

Worst-Case Scenario . . . Disaster Recovery

The need for adequate backup measures, both on- and off-site, is an out-of-the-mainstream lesson of 9/11. After the initial losses, many businesses found themselves in a worst-case scenario beyond their worst nightmares in terms of disaster recovery. Some estimates to replace the technology destroyed for securities firms affected by 9/11 placed the cost at approximately $3.2 billion: $1.7 billion spent on hardware—trading stations, sales stations, workstations, PCs, servers, printers, minicomputers, storage devices, cabling, communications hubs, routers, switches, and so on; $1.5 billion for software and services to install and connect the network, operating system, and software applications. And this is only for securities firms—not for all those affected.

Usually, disaster recovery planning provides desk space for only a business's more critical elements—usually about 20 percent of an organization's workforce. That means the other 80 percent of the workforce have no access to desks and computer systems. Getting the 80 percent to work again in the shortest possible time is essential for continuity.

Such planning was a hard sell before 9/11. Answering "what do I get for my money" was hard to justify, though many organizations had some reasonable level of disaster planning. Funding difficulties led to a two-tier approach: separate plans to cope with an immediate (short-term) incident (the disaster recovery plan) and longer-term major incident recovery methods (continuity plan).

After 9/11, when all of the affected organizations invoked their disaster recovery plans (in one way or another), the contracts with their disaster recovery service providers went into effect: emergency desk space previously put into a state of semireadiness were supposed to be available. The sheer scale of the disaster caused problems. These providers operate somewhat like insurance policies—they plan their operations by statistical analysis, and statistically, such massive failures are uncommon. While organizations had contracted for a specific space, the service providers had no way to provide for disaster on that scale—so many different companies invoking their contracts for use at once. Such companies usually provide a shared space with multiple companies contracting for the same space and using it at different times, as needed. After 9/11, everyone needed all of the space, at once, and not for just the usual couple of days or weeks expected after a company has problems. Many clients wanted the full use of their contract, for the full terms of their contracts—usually a maximum of six months.

Post 9/11 thinking for continuity planning involves the possibility of alternate premises, sometimes independent facilities (sometimes moving to another company site within the same organization) to allow the full staff a quick return to a normal workload.

Setting up a new installation from scratch takes time, often more than the six months most disaster recovery service contracts offer. While a client organization is heavily involved in restoring business as usual, the disaster recovery service provider must deal with many other clients who may need to share the same limited resources. The disaster recovery service suppliers end up seeking alternate space as well, possibly competing with their own clients for suitable housing (McCreary 2003).

Viral Protection

IT personnel should regularly update and share viral protection software as well as educate employees on the dangers of stranger-generated e-mail. As some folks have to climb a mountain "because it's there," other folks seem to feel obligated to trash computer systems on a widespread basis, either to prove they can or for financial gain. Computer server security systems involve solid firewall protection, but various Trojan horses, viruses, parasites, worms, and bugs are being created and spread all the time. A really massive infestation of some of the more wicked bugs can bring an organization to its knees for days and cost thousands of dollars of wasted work hours and ruined data and applications as well as hundreds of hours of IT staff time to fix. Make sure viral protection software is on all workstations and that it is regularly and completely updated. Ensure that anti-virus software is configured to identify viruses by all of each computer's means of entry—e-mail, web browsing, floppy disks, CDs, archives, and so on.

E-mail

E-mail is a wonderful tool for business and industry in many ways, but it is also one of the biggest security threats in many systems. E-mail systems demand effective management—one of the security policy's key elements. IDC (2009) predicted that by 2005, cyberspace and the physical components in people's offices and homes will be running 1.2 billion e-mail boxes (138 percent compounded growth) and 36 billion person-to-person e-mails daily. Instant messaging is a high-growth area, too—an estimated growth rate of 100 percent compounded, with 150 million business users by 2004.

Think about the number of e-mail-based or -delivered bug scares in the past year. Now consider what such growth may mean in terms of security problems. Virus defenses, e-mail data security, content filtering, and SPAM protection are key—and keeping them up to date cannot be ignored. Software solutions for e-mail protection help define and enforce an e-mail security policy and provide filtering, monitoring, and reporting. Effective software works in tandem with anti-virus software, providing additional protection company-wide, including protection designed to safeguard intellectual property and confidential information and methods of dealing with SPAM. Contextual searching picks up e-mails that might contain offensive materials and can determine whether employees are mailing out sensitive information such as customer lists or research data.

The Privacy Issue

The quick growth of the Internet and e-mail use has made employees' privacy rights versus corporate security/responsibility a pertinent and controversial issue. As e-mail and Internet monitoring grow, more and more disciplinary actions ensue, leaving individuals unsure about their rights to workplace privacy. The struggle to balance individual right to privacy and security with organizations' responsibility to protect their own intellectual assets has become increasingly intense.

As a middle ground between absolute freedom for employees to e-mail and surf at will and the "Big Brother sees all" effect of high levels of employee monitoring, a sensible and reasonable company policy offers compromise. Written company policy, effectively established, regularly discussed with employees, and consistently enforced through the appropriate technology is critical to maintaining consistency and predictability in the process of monitoring the workplace and helps address the privacy issue.

AUTOMATED TOOLS FOR DATA SECURITY SYSTEMS

Computer and data security systems must be suited to the needs of the company. Levels of protection should correspond to the value (both monetary and replacement) of the equipment and data of the business. Remember, however, that while automated tools do a great job in many areas, they can't do everything. Regular human-initiated checks must also be in place. Common automated tools for system protection include

- Firewall reporting—Firewall reports give complete accounts of everything that goes through a firewall and highlights attempts to breach security.
- Vulnerability testing—Vulnerability testing tools regularly test systems in different ways to illuminate system strengths and weaknesses.
- Intrusion detection—Intrusion detection identifies and responds rapidly to intrusion attempts.
- Web filtering—Limiting access to unneeded and unrelated websites lowers productivity costs generated by web abuse.
- Honey pot system—Decoy servers or systems set up to attract intrusion attention and gather information regarding an attacker or intruder are available for internal, external, and remote access systems.

Firewall

With the number of port scans carried out against computers while connected to the Internet, even single-user computers should have suitable firewall protection.

For personal use, simple software personal firewalls offer a reasonable degree of protection when correctly configured to the individual computer and operating system. Even with a firewall in place, open communication channels through the firewall can be exploited by hackers to gain access to information. Firewalls are even more critical in network situations. Sophisticated software/hardware combinations offer a higher level of protection for networks. Always-on connections offer excellent opportunities for hackers to gain access to systems and are also a route by which spam can be propagated, increasing the need for firewall protection.

Access Control

Log-on systems protect data on both networked workstations and stand-alone machines as well as laptops. Larger networks and companies should determine what needs protection and the level of data protection needed.

Internal Threats

While external threats are an ever-present problem in computer security these days, theft of data by a company employee is also a widespread problem, a form of theft made easier by the worldwide ease of information transmission. Any facility that uses or generates sensitive information should have systems in place to prevent internal theft. Industrial espionage is a reality, even if it probably isn't as sexy as it seems in books and movies. Limiting access to sensitive materials, encryption, special authorization processes for users, monitoring computer activity for those who handle sensitive materials, and blocks on electronic transmission of certain types of data are some ways to protect data. Clearly stated and specific company policy provides legal protection for the employer.

COMPUTER SECURITY . . . CRADLE TO GRAVE . . .

In our throwaway society, we sometimes forget that one man's trash is another man's treasure. Managing data on hard drives at the end of the PC life cycle is commonly ignored. However, data recovery is a growing and quickly expanding element in the computer industry. Data recovery companies can pull useable data off hard drives that were data cleansed, from laptops pulled from shipwrecks, from hard drives that went through fires or were otherwise severely physically abused. What can businesses do to ensure their old data is gone—and not available to someone who's just bought an old hard drive?

Several years ago, two students at the MIT Laboratory of Computer Science purchased 158 used hard drives with a scrap value of one thousand dollars. On some of those drives, they found five thousand credit card numbers and many medical reports, as well as corporate and financial information. Fortunately for the companies concerned, the students used this project as data in a report on their findings. In short, they concluded that someone buying any ten random drives on the used market would have a 30 percent chance of finding useful confidential information (Riches 2003).

How can a company destroy the information on redundant equipment—or at least protect it from recovery? Keeping such equipment forever isn't a great option—secure space is costly, and besides, the equipment is subject to theft. Be sure data removal tools are used on redundant equipment, and remember that the distance IT should go to protect that data should depend on the value of that data to outside sources.

IN-HOUSE OR OUTSOURCED SECURITY?

Just as outsourcing emergency services is more difficult to achieve than it sounds, outsourcing security poses a unique set of problems. Outsourced computer system security is bound to be expensive and, as with other safety needs, has no direct financial benefit to the organization in the eyes of most financial officers. Security is an insurance policy—those financial officers will have very little way to know what any security system has saved them—unless the savings are detailed to them. When determining whether computer and data security should be accomplished in-house or outsourced, estimating the costs of both is essential.

Determining the level of expertise already available in-house is the next consideration. If local talent isn't capable of the job, the cost of new hires must be figured in. While lack of internal recourses might make outsourcing seem more attractive, the element of trust is higher for in-house systems. Security is usually outsourced when company resources can't handle it. Realize that outsourcing cannot be accomplished without in-house involvement at some level. Outsourcing does not remove the ultimate security responsibility from your organization—so the question becomes "will my company computer system be more secure with in-house or outsourced security?"

Outsourcing concerns include return on investment considerations, risk assessment, security spending versus value of the data to be protected, and the possible consequences of a security breach.

One of the most important outsource concerns is training. Who will be responsible for staff training? Will training be in-house or through outside consultants? How much will training and staff education cost—for both in-house and outsourced? Will you run seminars? Will you produce your own study materials?

What to ask when buying security (Franklin 2003):

- Do you know what you need to protect?
- Do you know what you are protecting it from?
- Have you completed a risk assessment?
- Have you researched the legal obligations of your company?
- Does your security strategy mirror your business strategy?
- Have you considered the training implications for users?

- Do you have appropriate skills and resources in-house to fulfill the project, or do you need outside help?
- Have you planned for the use of mobile technology within the organization?
- Have you thoroughly investigated all the options for both in-house or outsourced security?

CYBER SECURITY: THE BOTTOM LINE

The need for computer security is constantly growing, expanding, and changing. Because the computer industry as a whole is experiencing such levels of growth, the issues safety and security engineers face today may well be different tomorrow. Only by establishing a solid working relationship with IT can even a fully computer-savvy safety and security engineer have any hope of keeping abreast of this critical area. Regular research is essential to maintaining the integrity of computer systems and data.

REFERENCES AND RECOMMENDED READING

A guide to rolling out handhelds in the corporate environment. 2009. http://www.itsecurity .com/papers/psion1.htm.

Berinato, S. 2003. *The future of security.* http://www.computerworld.com/s/article/print/ 88646/The_future_of_security (accessed December 25, 2009).

Cox, Peter. 2009. *Security evaluation: The common criteria certifications.* http://www.itsecurity .com/papers/border.htm.

CRS. 2003a. *Computer attack and cyber terrorism: Vulnerabilities and policy issues for Congress.* Clay Wilson. Washington, DC: Library of Congress, Congressional Research Service.

———. 2003b. *Information warfare and cyberwar: Capabilities and related policy issues.* Clay Wilson. Washington, DC: Library of Congress, Congressional Research Service.

Danchev, D. 2003. *Reducing "human factor" mistakes.* http://www.itsecurity.com/papers/ dancho1.htm.

Dick, R. 2002. *The truth about cyberterrorism.* Washington, DC: NIPC (DHS).

Ellison, R. J., et al. 1997. *An approach to survivable systems.* Pittsburgh: CERT Coordination Center, Software Engineering Institute, Carnegie Mellon University.

Franklin, I. 2003. *How to buy security.* http://www.itsecurity.com/papers/entercept3.htm.

IDC. 2009. *Email usage forecast and analysis, 2000–2005.* Prnewswire.com/cgi-bin/stories. pl?ACCT:104&STORY=/ . . . &EDATA (accessed December 24, 2009).

Kilpatrick, I. 2002. *Security—The challenges of today And tomorrow.* http://www.itsecurity .com/papers/wick3.htm.

———. 2003. *Security patching: Are you guilty of corporate stupidity?* http://www.itsecurity .com/papers/wick5.htm.

Mathews, S. 2002. *Security can be simple and secure.* http://www.itsecurity.com/papers/ articsoft15.htm.

McCreary, L. 2003. *A proven paradigm for best practices in information security.* http://www .itsecurity.com/papers/securecomp2.htm.

Millard, Nathan. 2003. *Achieving policy management best practice: A guide to improving corporate governance and realizing more value from policies.* http://www.itsecurity.com/ papers/extend1.htm.

NRC. 2006. *Regulatory guide 1.1.52: Criteria for use of computers in safety systems of nuclear power plants.* Washington, DC: Nuclear Regulatory Commission.

———. 2009. Protection of digital computer and communication systems and networks. http://nrc.gov/reading-rm/doc-collections/cfr/part073/part073-0054.html (accessed December 22, 2009).

Rasiah, S. 2002. *Putting a price on leaked information.* http://www.itsecurity.com/papers/ pentasafe2.htm.

Riches, K. 2003. *Data protection.* http://www.itsecurity.com/papers/tam1.htm.

Rose, C. 2002. *Computer security audit checklist.* http://www.itsecurity.com/papers/iomart2 .htm.

Royce, W. 1970. Managing the development of large software systems. *Proceedings of IEEE WESCON* 26 (August): 1–9.

SANS Institute. 2002–2003. *Terminology.* http://www.sans.org/resources/glossary.php.

Shore, D. 2002. *Lessons emerging from disaster.* http://www.itsecurity.com/papers/shore1.htm.

Skinner, M. 2002. *Tips for information security.* http://www.itsecurity.com/papers/axent.htm.

Skoudis, E. 2002. *Counter hack.* Upper Saddle River, NJ: Prentice Hall.

Tickle, I. 2002. *Data integrity—The unknown threat.* http://www.itsecurity.com/papers/ tripwire1.htm.

Training and awareness programme. http://www.itsecurity.com/papers/trinity8.htm.

Trinity Security Services. 2003. *Desktop security.* http://www.itsecurity.com/papers/trinity13 .htm.

8

SCADA

Unless people are injured, there is also less drama and emotional appeal.

—*Dorothy Denning (November 1, 2001)*

On April 23, 2000, police in Queensland, Australia stopped a car on the road and found a stolen computer and radio inside. Using commercially available technology, a disgruntled former employee had turned his vehicle into a pirate command center of sewage treatment along Australia's Sunshine Coast. The former employee's arrest solved a mystery that had troubled the Maroochy Shire wastewater system for two months. Somehow the system was leaking hundreds of thousands of gallons of putrid sewage into parks, rivers and the manicured grounds of a Hyatt Regency hotel—marine life died, the creek water turned black and the stench was unbearable for residents. Until the former employee's capture—during his 46th successful intrusion—the utility's managers did not know why.

Specialists study this case of cyber-terrorism because it is the only one known in which someone used a digital control system deliberately to cause harm. The former employee's intrusion shows how easy it is to break in—and how restrained he was with his power.

To sabotage the system, the former employee set the software on his laptop to identify itself as a pumping station, and then suppressed all alarms. The former employee was the "central control station" during his intrusions, with unlimited command of 300 SCADA nodes governing sewage and drinking water alike.

The bottom line: as serious as the former employee's intrusions were they pale in comparison with what he could have done to the fresh water system—he could have done anything he liked.

—*Barton Gellman, 2002*

IN THE WORDS OF MASTER SUN TZU,
FROM *THE ART OF WAR*

Those who are first on the battlefield, and await the opponents are at
ease; those who are last, and head into battle are worn out.

In 2000, the FBI identified and listed threats to critical infrastructure. These threats are
listed and described in table 8.1.

Table 8.1. Threats to Critical Infrastructure Observed by the FBI

Threat	Description
Criminal groups	There is an increased use of cyber intrusions by criminal groups who attack systems for purposes of monetary gain.
Foreign intelligence services	Foreign intelligence services use cyber tools as part of their information-gathering and espionage activities.
Hackers	Hackers sometimes crack into networks for the thrill of the challenge or for bragging rights in the hacker community. While remote cracking once required a fair amount of skill or computer knowledge, hackers can now download attack scripts and protocols from the Internet and launch them against victim sites. Thus, while attack tools have become more sophisticated, they have also become easier to use.
Hacktivists	Hacktivism refers to politically motivated attacks on publicly accessible web pages or e-mail servers. These groups and individuals overload e-mail servers and hack into websites to send a political message.
Information warfare	Several nations are aggressively working to develop information warfare doctrine, programs, and capabilities. Such capabilities enable a single entity to have a significant and serious impact by disrupting the supply, communications, and economic infrastructures that support military power—impacts that, according to the Director of Central Intelligence, can affect the daily lives of Americans across the country.
Inside threat	The disgruntled organization insider is a principal source of computer crimes. Insiders may not need a great deal of knowledge about computer intrusions because their knowledge of a victim system often allows them to gain unrestricted access to cause damage to the system or to steal system data. The insider threat also includes outsourcing vendors.
Virus writers	Virus writers are posing an increasingly serious threat. Several destructive computer viruses and worms have harmed files and hard drives, including the Melissa Macro Virus, the Explore.Zip worm, the CIH (Chernobyl) Virus, Nimda, and Code Red.

Source: FBI, *Threat to Critical Infrastructure* (Washington, DC: Federal Bureau of Investigation, 2000).

NUCLEAR POWER AND CYBERSPACE

As pointed out in chapter 7, today's developing "information age" technology has intensified the importance of critical infrastructure protection, in which cyber security has become as critical as physical security to protecting all sectors of critical infrastructure.

In the past few years, especially since 9/11, it has been somewhat routine for us to pick up a newspaper or magazine or view a television news program where a major topic of discussion is cyber security or the lack thereof. Many of the cyber intrusion incidents we read or hear about have added new terms or new uses for old terms to our vocabulary. For example, old terms such as *botnets* (short for robot networks, also balled bots, zombies, botnet fleets, and many others) are groups of computers that have been compromised with malware such as Trojan horses, worms, backdoors, remote control software, and viruses have taken on new connotations in regard to cyber security issues. Relatively new terms such as scanners, Windows NT hacking tools, ICQ hacking tools, mail bombs, sniffers, logic bombs, nukers, dots, backdoor Trojans, key loggers, hackers' Swiss knife, password crackers, blended threats, Warhol worms, Flash threats, targeted attacks, and BIOS crackers are now commonly read or heard. New terms have evolved along with various control mechanisms. For example, because many control systems are vulnerable to attacks of varying degrees, these attack attempts range from telephone line sweeps (war dialing), to wireless network sniffing (war driving), to physical network port scanning, to physical monitoring and intrusion. When wireless network sniffing is performed at (or near) the target point by a pedestrian (war walking), meaning that instead of a person being in an automotive vehicle, the potential intruder may be sniffing the network for weaknesses or vulnerabilities on foot, posing as a person walking, but the person may have a handheld PDA device or laptop computer (Warwalking 2003).

Not all relatively new and universally recognizable cyber terms have sinister connotation or meaning, of course. Consider, for example, the following digital terms: backup, binary, bit, byte, CD-ROM, CPU, database, e-mail, HTML, icon, memory, cyberspace, modem, monitor, network, RAM, Wi-Fi (wireless fidelity), record, software, World Wide Web—none of these terms normally generate thoughts of terrorism in most of us.

There is, however, one digital term, SCADA—first used in the 1960s—that most people have not heard of. This is not the case, however, with those who work with the nation's critical infrastructure, including nuclear power plant sector infrastructure. SCADA, or Supervisory Control And Data Acquisition system (also sometimes referred to as digital control system or process control system), plays an important role in computer-based control systems. From coordinating music and lights in the proper sequence with spray from water fountains to controlling systems used in the drilling and refining of oil and natural gas, control systems perform many functions. Other examples include energy distribution networks that use computer-based systems to remotely control sensitive feeds, processes, and system equipment previously controlled manually. These systems (commonly known as SCADA) allow a critical infrastructure sector entity (or operation) to monitor fuel tank levels, to ensure that contents are stored at correct levels or to monitor tank levels, and, as mentioned, to collect data from sensors and control equipment located at remote sites. Common critical infrastructure sector system sensors monitor and measure elements such as fluid level, temperature, pressure, petrochemical purity or slurry composition, petrochemical purity, and energy and petrochemical pipeline flow rates, based on feedback data gathered by sensors. Common critical infrastructure industry system equipment includes valves, pumps, and switching devices for distribution of electricity. SCADA systems can be relatively simple, such as one that monitors environmental conditions of an office building, or incredibly complex, such as a system that monitors the activity of a municipal water system or all the activity in a nuclear power plant.

The critical infrastructure of many countries is increasingly dependent on SCADA systems. For this reason, some experts believe that these systems may be vulnerable to cyber attack and that their importance for controlling the critical infrastructure may make them an attractive target for cyberterrorists. SCADA systems once used only proprietary computer software, and their operation was confined largely to isolated networks. However, an increasing number of industrial control systems now operate using commercial off-the-shelf (COTS) software, and more are being linked by the Internet directly into their corporate headquarters office systems. Some industry experts believe that SCADA systems are inadequately protected against a cyber attack and remain vulnerable because many of the organizations that operate them have not paid proper attention to computer security needs (CRS 2003).

KEEP IN MIND

Stealth and preoperational surveillance are important characteristics known to precede a computer attack launched by hackers. Similar characteristics have also been described as a "hallmark" of some previous al Qaeda physical terrorist attacks and bombings (CRS 2003).

WHAT IS SCADA?

If we were to ask the specialist, "What is SCADA?" the technical response could be outlined as follows:

SCADA is

- A multitier system or interfaces with multitier systems
- Used for physical measurement and control endpoints via an RTU and PLC to measure voltage, adjust a value, or flip a switch
- An intermediate processor normally based on commercial third-party OSes—VMS, Unix, Windows, Linux
- A means of human interface, for example, with a graphical user interface (Windows GUIs)
- A communication infrastructure consisting of a variety of transport mediums such as analog, serial, Internet, radio, and Wi-Fi

How about the nonspecialist response—for the rest of us who are nonspecialists?

Well, for the nonspecialist and the rest of us, we could simply say that SCADA is a computer-based control system that remotely controls processes previously controlled manually. The philosophy behind SCADA control systems can be summed up by the phrase "If you can measure it, you can control it." SCADA allows an operator using a central computer to supervise (control and monitor) multiple networked computers at remote locations. Each remote computer can control mechanical processes (mixers, pumps, valves, etc.) and collect data from sensors at its remote location. Thus the phrase Supervisory Control and Data Acquisition, or **SCADA**.

The central computer is called the master terminal unit, or MTU. The MTU has two main functions: periodically obtain data from RTUs/PLCs, and control remote devices through the operator station. The operator interfaces with the MTU using software called human machine interface, or HMI. The remote computer is called the program logic controller (PLC) or remote terminal unit (RTU). The RTU activates a relay (or switch) that turns mechanical equipment "on" and "off." The RTU also collects data from sensors. Sensors perform measurement, and actuators perform control.

In the initial stages, utilities ran wires, also known as hardwire or land lines, from the central computer (MTU) to the remote computers (RTUs). Since remote locations can be located hundreds of miles from the central location, utilities began to use public phone lines and modems, leased telephone company lines, and radio and microwave communication. More recently, they have also begun to use satellite links, the Internet, and newly developed wireless technologies.

DID YOU KNOW?

Modern RTUs typically use a ladder-logic approach to programming due to its similarity to standard electrical circuits. An RTU that employees this ladder-logic programming is called a programmable logic controller (PLC).

Because SCADA systems' sensors provided valuable information, many utilities and other industries established "connections" between their SCADA systems and their business system. This allowed utility/industrial management and other staff access to valuable statistics, such as chemical usage. When utilities/industries later connected their systems to the Internet, they were able to provide stakeholders/stockholders with usage statistics on the utility/industrial web pages. Figure 8.1 provides a basic illustration of a representative SCADA network. Note that firewall protection would normally be placed between Internet and business system and between business system and the MTU.

SCADA Applications in Nuclear Power Sector Systems

As stated above, SCADA systems can be designed to measure a variety of equipment operating conditions and parameters or volumes and flow rates or electricity, natural gas, and oil or oil and petrochemical mixture quality parameters, and to respond to change in those parameters either by alerting operators or by modifying system operation through a feedback loop system without having personnel physically visit each process or piece of equipment on a daily basis to check it and/or ensure that it is functioning properly. Automation and integration of large-scale diverse assets required SCADA systems to provide the utmost in flexibility, scalability, openness, and reliability. SCADA systems are used to automate certain nuclear power energy production functions; these can be performed without initiation by an operator. In addition to process equipment, SCADA systems can also integrate specific security alarms and equipment, such as cameras, motion sensors, lights, data from card-reading systems, and so on, thereby providing a clear picture of what is happening at areas throughout a facility. Finally, SCADA systems also provide constant, real-time data on processes, equipment, location access, and so on, the necessary response to be made quickly. This can be extremely useful during emergency conditions, such as when energy distribution lines or piping breaks or when potentially disruptive spikes appear in plant processing operations.

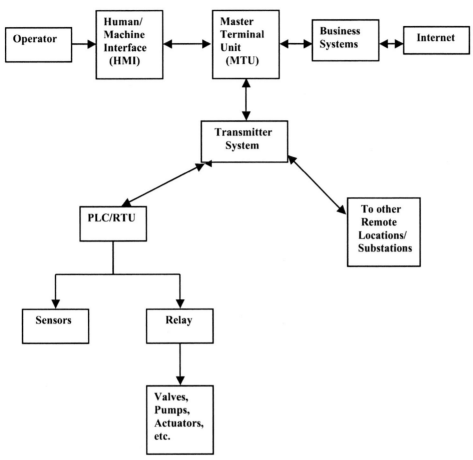

FIGURE 8.1
Representative SCADA Network

Today, common nuclear power sector applications for SCADA systems include, but are not limited to, those listed below.

- Bearing temperature monitor (electric generators and motors)
- Gas processing
- Plant energy management
- Power distribution monitoring
- Electric power T&D monitoring
- Plant monitoring
- Process controls
- Process stimulators

- Tank controls
- Utility monitoring
- Safety parameter display systems
- Turbine controls
- Turbine monitoring
- Virtual annunciator panels

Because these systems can monitor multiple processes, equipment, and infrastructure and then provide quick notification of, or response to, problems or upsets, SCADA systems typically provide the first line of detection for atypical or abnormal conditions. For example, a SCADA system connected to sensors that measure specific nuclear power or plant support equipment quality parameters are measured outside of a specific range. A real-time customized operator interface screen could display and control critical systems monitoring parameters.

The system could transmit warning signals back to the operators, such as by initiating a call to a personal pager. This might allow the operators to initiate actions to prevent power outages or contamination and disruption of the energy supply. Further automation of the system could ensure that the system initiated measures to rectify the problem. Preprogrammed control functions (e.g., shutting a valve, controlling flow, throwing a switch, or adding chemicals) can be triggered and operated based on SCADA utility.

SCADA VULNERABILITIES

U.S. Electric Grid Gets Hacked Into
The Associated Press (AP) reported April 9, 2009, that spies hacked into the U.S. energy grid and left behind computer programs (Trojan horses) that would enable them to disrupt service, exposing potentially catastrophic vulnerabilities in key pieces of national infrastructure.

Slammer Worm Crashed Ohio Nuke Plant Network
In 2003, the "Slammer" worm corrupted for 5 hours the computer systems at the closed Davis-Besse nuclear power plant located in Ohio. The worm bypassed firewall security, and highlighted possible security issues that may arise whenever plant networks and corporate networks are interconnected. The Davis-Besse corporate network was found to have multiple connections to the Internet that bypassed the plant firewall. (Poulsen 2003)

Even though terrorists, domestic and/or foreign, tend to aim their main focus around the critical devices that control actual critical infrastructure sectors, such as energy production (including nuclear power–generated energy production) and delivery, according to U.S. EPA (2005) and mentioned earlier, SCADA networks were developed with little attention paid to security, making the security of these systems often weak. Studies have found that, while technological advancements introduced vulnerabilities, many critical infrastructure sector plans/sites and utilities have spent little time securing their SCADA networks. As a result, many SCADA networks may be susceptible to attacks and misuse. SCADA systems languished in obscurity, and this was the essence of their security; that is, until technological developments transformed SCADA from a backroom operation to a front-and-center visible control system.

Remote monitoring and supervisory control of processes began to develop in the early 1960s and adopted many technological advancements. The advent of minicomputers made it possible to automate a vast number of once manually operated switches. Advancements in radio technology reduced the communication costs associated with installing and maintaining buried cable in remote areas. SCADA systems continued to adopt new communication methods, including satellite and cellular. As the price of computers and communications dropped, it became economically feasible to distribute operations and to expand SCADA networks to include even smaller facilities.

Advances in information technology and the necessity of improved efficiency have resulted in increasingly automated and interlinked infrastructures and created new vulnerabilities due to equipment failure, human error, weather and other natural causes, and physical and cyber attacks. Some areas and examples of possible SCADA vulnerabilities include

- Human—People can be tricked or corrupted and may commit errors.
- Communications—Messages can be fabricated, intercepted, changed, deleted, or blocked.
- Hardware—Security features are not easily adapted to small self-contained units with limited power supplies.
- Physical—Intruders can break into a facility to steal or damage SCADA equipment.
- Natural—Tornadoes, floods, earthquakes, and other natural disasters can damage equipment and connections.
- Software—Programs can be poorly written.

As a case in point related to actions not properly taken in securing SCADA, consider a study that included a survey that found that many water utilities were doing little to secure their SCADA network vulnerabilities (Ezell 1998). For example, many respondents reported that they had remote access, which can allow an unauthorized person to access the system without being physically present. More than 60 percent of the respondents believed that their systems were not safe from unauthorized access and use. Twenty percent of the respondents even reported known attempts, successful unauthorized access, or use of their system. Yet twenty-two of forty-three respondents reported that they do not spend any time ensuring their network is safe, and eighteen of forty-three respondents reported that they spend less than 10 percent ensuring network safety.

SCADA system computers and their connections are susceptible to different types of information system attacks and misuse, such as system penetration and unauthorized access to information. The Computer Security Institute and Federal Bureau of Investigation conduct an annual Computer Crime and Security Survey (FBI 2004). The survey reported on ten types of attacks or misuse and reported that virus and denial of service had the greatest negative economic impact. The same study also found that 15 percent of the respondents reported abuse of wireless networks, which can be a SCADA component. On average, respondents from all sectors did not believe that their organization invested enough in security awareness. Utilities as a group reported a lower average computer security expenditure/investment per employee than many other sectors, such as transportation, telecommunications, and financial.

Sandia National Laboratories' *Common Vulnerabilities in Critical Infrastructure Control Systems* described some of the common problems it has identified in the following five categories (Stamp et al. 2003):

1. **System Data**—Important data attributes for security include availability, authenticity, integrity, and confidentiality. Data should be categorized according to its sensitivity, and ownership and responsibility must be assigned. However, SCADA data is often not classified at all, making it difficult to identify where security precautions are appropriate (for example, which communication links to secure, databases requiring protection, etc.).
2. **Security Administration**—Vulnerabilities emerge because many systems lack a properly structured security policy (security administration is notoriously lax in the case of control systems), equipment and system implementation guides, configuration management, training, and enforcement and compliance auditing.
3. **Architecture**—Many common practices negatively affect SCADA security. For example, while it is convenient to use SCADA capabilities for other purposes such as fire and security systems, these practices create single points of failure. Also, the connection of SCADA networks to other automation systems and business networks introduces multiple entry points for potential adversaries.
4. **Network** (including communication links)—Legacy systems' hardware and software have very limited security capabilities, and the vulnerabilities of contemporary systems (based on modern information technology) are publicized. Wireless and shared links are susceptible to eavesdropping and data manipulation.

5. **Platforms**—Many platform vulnerabilities exist, including default configurations retained, poor password practices, shared accounts, inadequate protection for hardware, and nonexistent security monitoring controls. In most cases, important security patches are not installed, often due to concern about negatively impacting system operation; in some cases technicians are contractually forbidden from updating systems by their vendor agreements.

The following incident helps to illustrate some of the risks associated with SCADA vulnerabilities.

- During the course of conducting a vulnerability assessment, a contractor stated that personnel from his company penetrated the information system of a utility within minutes. Contractor personnel drove to a remote substation and noticed a wireless network antenna. Without leaving their vehicle, they plugged in their wireless radios and connected to the network within five minutes. Within twenty minutes they had mapped the network, including SCADA equipment, and accessed the business network and data. This illustrates what a cyber security adviser from Sandia National Laboratories specializing in SCADA stated, that utilities are moving to wireless communication without understanding the added risks.

The Increasing Risk

According to GAO (2003), historically, security concerns about control systems (SCADA included) were related primarily to protecting against physical attack and misuse of refining and processing sites or distribution and holding facilities. However, more recently there has been a growing recognition that control systems are now vulnerable to cyber attacks from numerous sources, including hostile governments, terrorist groups, disgruntled employees, and other malicious intruders.

In addition to control system vulnerabilities mentioned earlier, several factors have contributed to the escalation of risk to control systems, including (1) the adoption of standardized technologies with known vulnerabilities, (2) the connectivity of control systems to other networks, (3) constraints on the implementation of existing security technologies and practices, (4) insecure remote connections, and (5) the widespread availability of technical information about control systems.

Adoption of Technologies with Known Vulnerabilities

When a technology is not well known, not widely used, and not understood or publicized, it is difficult to penetrate it and thus disable it. Historically, proprietary hardware, software, and network protocols made it difficult to understand how control systems operated—and therefore how to hack into them. Today, however, to reduce costs and improve performance, organizations have been transitioning from proprietary systems to less expensive, standardized technologies such as Microsoft's Windows and Unix-like operating systems, and the common networking protocols used by the Internet. These widely used standardized technologies have commonly known vulnerabilities, and sophisticated and effective exploitation tools are widely available and relatively easy to use. As a consequence, both the number of people with the knowledge to wage attacks and the number of systems subject to attack have increased. Also, common communication protocols and the emerging use of Extensible Markup Language (commonly referred to as XML) can make it easier for a hacker to interpret the content of communications among the components of a control system.

Control systems are often connected to other networks—enterprises often integrate their control system with their enterprise networks. This increased connectivity has significant advantages, including providing decision makers with access to real-time information and allowing engineers to monitor and control the process control system from different points on the enterprise network. In addition, the enterprise networks are often connected to the networks of strategic partners and to the Internet. Further, control systems are increasingly using wide area networks and the Internet to transmit data to their remote or local stations and individual devices. This convergence of control networks with public and enterprise networks potentially exposes the control systems to additional security vulnerabilities. Unless appropriate security controls are deployed in the enterprise network and the control system network, breaches in enterprise security can affect the operation of control systems.

According to industry experts, the use of existing security technologies, as well as strong user authentication and patch management practices, are generally not implemented in control systems because control systems operate in real time, typically are not designed with cyber security in mind, and usually have limited processing capabilities.

Existing security technologies such as authorization, authentication, encryption, intrusion detection, and filtering of network traffic and communications require more bandwidth, processing power, and memory than control system components typically have. Because controller stations are generally designed to do specific tasks, they use low-cost, resource-constrained microprocessors. In fact, some devices in the electrical industry still use the Intel 8088 processor, introduced in 1978. Consequently, it is difficult to install existing security technologies without seriously degrading the performance of the control system.

Further, complex passwords and other strong password practices are not always used to prevent unauthorized access to control systems, in part because this could hinder a rapid response to safety procedures during an emergency. As a result, according to experts, weak passwords that are easy to guess, shared, and infrequently change are reportedly common in control systems, including the use of default passwords or even no password at all.

In addition, although modern control systems are based on standard operating systems, they are typically customized to support control system applications. Consequently, vendor-provided software patches are generally either incompatible or cannot be implemented without compromising service shutting down "always-on" systems or affecting interdependent operations.

Potential vulnerabilities in control systems are exacerbated by insecure connections. Organizations often leave access links—such as dial-up modems to equipment and control information—open for remote diagnostics, maintenance, and examination of system status. Such links may not be protected with authentication or encryption, which increases the risk that hackers could use these insecure connections to break into remotely controlled systems. Also, control systems often use wireless communications systems, which are especially vulnerable to attack, or leased lines that pass through commercial telecommunications facilities. Without encryption to protect data as it flows through these insecure connections or authentication mechanisms to limit access, there is limited protection for the integrity of the information being transmitted.

Public information about infrastructures and control systems is available to potential hackers and intruders. The availability of this infrastructure and vulnerability data was demonstrated by a university graduate student, whose dissertation reportedly mapped every business and industrial sector in the American economy to the fiber-optic network that connects them—using material that was available publicly on the Internet, none of which was classified. Many of the electric utility officials who were interviewed for the National Security Telecommunications Advisory Committee's Information Assurance Task Force's Electric Power Risk Assessment expressed concern over the amount of information about their infrastructure that is readily available to the public.

In the electric power industry (including that generated by nuclear power plants), open sources of information—such as product data and educational videotapes from engineering associations—can be used to understand the basics of the electrical grid. Other publicly available information—including filings of the Federal Energy Regulatory Commission (FERC), industry publications, maps, and material available on the Internet—is sufficient to allow someone to identify the most heavily loaded transmission lines and the most critical substations in the power grid.

In addition, significant information on control systems is publicly available—including design and maintenance documents, technical standards for the interconnection of control systems and RTUs, and standards for communication among control devices—all of which could assist hackers in understanding the systems and how to attack them. Moreover, there are numerous former employees, vendors, support contractors, and other end users of the same equipment worldwide with inside knowledge of the operation of control systems.

Cyber Threats to Control Systems

There is a general consensus—and increasing concern—among government officials and experts on control systems about potential cyber threats to the control systems that govern our critical infrastructures. As components of control systems increasingly make critical decisions that were once made by humans, the potential effect of a cyber threat becomes more devastating. Such cyber threats could come from numerous sources, ranging from hostile governments and terrorist groups to disgruntled employees and other malicious intruders. Based on interviews and discussions with representatives throughout the electric power industry, the Information Assurance Task Force of the National Security Telecommunications Advisory Committee concluded that an organization with sufficient resources, such as a foreign intelligence service or a well-supported terrorist group, could conduct a structured attack on the electric power grid electronically, with a high degree of anonymity and without having to set foot in the target nation.

In July 2002, National Infrastructure Protection Center (NIPC) reported that the potential for compound cyber and physical attacks, referred to as "swarming attacks," is an emerging threat to the U.S. critical infrastructure. As NIPC reports, the effects of a swarming attack include slowing or complicating the response to a physical attack. For instance, a cyber attack that disabled the water supply or the electrical system in conjunction with a physical attack could deny emergency services the necessary resources to manage the consequences—such as controlling fires, coordinating actions, and generating light.

Control systems, such as SCADA, can be vulnerable to cyber attacks. Entities or individuals with malicious intent might take one or more of the following actions to successfully attack control systems:

- Disrupt the operation of control systems by delaying or blocking the flow of information through control networks, thereby denying availability of the networks to control system operations.
- Make unauthorized changes to programmed instructions in PLCs, RTUs, or DCS controllers, change alarm thresholds, or issue unauthorized commands to control equipment, which could potentially result in damage to equipment (if tolerances are exceeded), premature shutdown of processes (such as prematurely shutting down transmission lines), or even disabling of control equipment.
- Send false information to control system operators either to disguise unauthorized changes or to initiate inappropriate actions by system operators.

- Modify the control system software, producing unpredictable results.
- Interfere with the operation of safety systems.

In addition, in control systems that cover a wide geographic area, the remote sites are often unstaffed and may not be physically monitored. If such remote systems are physically breached, the attackers could establish a cyber connection to the control network.

Securing Control Systems

Several challenges must be addressed to effectively secure control systems against cyber threats. These challenges include (1) the limitations of current security technologies in securing control systems, (2) the perception that securing control systems may not economically justifiable, and (3) the conflicting priorities within organizations regarding the security of control systems.

A significant challenge in effectively securing control systems is the lack of specialized security technologies for these systems. The computing resources in control systems that are needed to perform security functions tend to be quite limited, making it very difficult to use security technologies within control system networks without severely hindering performance.

Securing control systems may not be perceived as economically justifiable. Experts and industry representatives have indicated that organizations may be reluctant to spend more money to secure control systems. Hardening the security of control systems would require industries to expend more resources, including acquiring more personnel, providing training for personnel, and potentially prematurely replacing current systems that typically have a life span of about twenty years.

Finally, several experts and industry representatives indicate that the responsibility for securing control systems typically includes two separate groups: IT security personnel and control system engineers and operators. IT security personnel tend to focus on securing enterprise systems, while control system engineers and operators tend to be more concerned with the reliable performance of their control systems. Further, they indicate that as a result, those two groups do not always fully understand each other's requirements and collaborate to implement secure control systems.

STEPS TO IMPROVE SCADA SECURITY

The President's Critical Infrastructure Protection Board and the Department of Energy (DOE) have developed the steps outlined below to help organizations improve the security of their SCADA networks. DOE (2001) points out that these steps are not meant to be prescriptive or all inclusive. However, they do address essential actions to be taken to improve the protection of SCADA networks. The steps are divided into two categories: specific actions to improve implementation, and actions to establish essential underlying management processes and policies.

Twenty-One Steps to Increase SCADA Security (DOE 2001)

The following steps focus on specific actions to be taken to increase the security of critical infrastructure SCADA networks:

1. **Identify all connections to SCADA networks.**
   Conduct a thorough risk analysis to assess the risk and necessity of each connection to the SCADA network. Develop a comprehensive understanding of all connections to the SCADA network and how well those connections are protected. Identify and evaluate the following types of connections:
   - Internal local area and wide area networks, including business networks
   - The Internet
   - Wireless network devices, including satellite uplinks
   - Modem or dial-up connections
   - Connections to business partners, vendors, or regulatory agencies
2. **Disconnect unnecessary connections to the SCADA network.**
   To ensure the highest degree of security of SCADA systems, isolate the SCADA network from other network connections to as great a degree as possible. Any connection to another network introduces security risks, particularly if the connection creates a pathway from or to the Internet. Although direct connections with other networks may allow important information to be passed efficiently and conveniently, insecure connections are simply not worth the risk; isolation of the SCADA network must be a primary goal to provide needed protection. Strategies such as utilization of "demilitarized zones" (DMZs) and data warehousing can facilitate the secure transfer of data from the SCADA network to business networks. However, they must be designed and implemented properly to avoid introduction of additional risk through improper configuration.
3. **Evaluate and strengthen the security of any remaining connections to the SCADA networks.**
   Conduct penetration testing or vulnerability analysis of any remaining connections to the SCADA network to evaluate the protection posture associated with these pathways. Use this information in conjunction with risk management processes to develop a robust protection strategy for any pathways to the SCADA network. Since the SCADA network is only as secure as its weakest connecting point, it is essential to implement firewalls, intrusion detection systems (IDSs), and other appropriate security measures at each point of entry. Configure firewall rules to prohibit access from and to the SCADA network, and be as specific

as possible when permitting approved connections. For example, an Independent System Operator (ISO) should not be granted "blanket" network access simply because there is a need for a connection to certain components of the SCADA system. Strategically place IDSs at each entry point to alert security personnel of potential breaches of network security. Organization management must understand and accept responsibility or risks associated with any connection to the SCADA network.

4. **Harden SCADA networks by removing or disabling unnecessary services.**
   SCADA control servers built on commercial or open-source operating systems can be exposed to attack through default network services. To the greatest degree possible, remove or disable unused services and network daemons to reduce the risk of direct attack. This is particularly important when SCADA networks are interconnected with other networks. Do not permit a service or feature on a SCADA network unless a thorough risk assessment of the consequences of allowing the service/feature shows that the benefits of the service/feature far outweigh the potential for vulnerability exploitation. Examples of services to remove from SCADA networks include automated meter reading/remote billing systems, e-mail services, and Internet access. An example of a feature to disable is remote maintenance. Use numerous secure configurations such as the National Security Agency's series of security guides. Additionally, work closely with SCADA vendors to identify secure configurations and coordinate any and all changes to operational systems to ensure that removing or disabling services does not cause downtime, interruption of service, or loss of support.

5. **Do not rely on proprietary protocols to protect your system.**
   Some SCADA systems are unique, proprietary protocols for communications between field devices and servers. Often the security of SCADA systems is based solely on the secrecy of these protocols. Unfortunately, obscure protocols provide very little "real" security. Do not rely on proprietary protocols or factory default configuration settings to protect your system. Additionally, demand that vendors disclose any backdoors or vendor interfaces to your SCADA systems, and expect them to provide systems that are capable of being secured.

6. **Implement the security features provided by device and system vendors.**
   Older SCADA systems (most systems in use) have no security features whatsoever. SCADA system owners must insist that their system vendor implement security features in the form of product patches or upgrades. Some newer SCADA devices are shipped with basic security features, but these are usually disabled to ensure ease of installation. Analyze each SCADA device to determine whether security features are present. Additionally, factory default security settings (such as in computer network firewalls) are often set to provide maximum usability but

minimal security. Set all security features to provide the maximum security only after a thorough risk assessment of the consequences of reducing the security level.

7. **Establish strong controls over any medium that is used as a backdoor into the SCADA network.**

Where backdoors or vendor connections do exist in SCADA systems, strong authentication must be implemented to ensure secure communications. Modems, wireless, and wired networks used for communications and maintenance represent a significant vulnerability to the SCADA network and remote sites. Successful war dialing or war driving attacks could allow an attacker to bypass all other controls and have direct access to the SCADA network or resources. To minimize the risk of such attacks, disable inbound access and replace it with some type of callback system.

8. **Implement internal and external intrusion detection systems and establish twenty-four-hour-a-day incident monitoring.**

To be able to effectively respond to cyber attacks, establish an intrusion detection strategy that includes alerting network administrators of malicious network activity originating from internal or external sources. Intrusion detection system monitoring is essential twenty-four hours a day; this capability can be easily set up through a pager. Additionally, incident response procedures must be in place to allow an effective response to any attack. To complement network monitoring, enable logging on all systems and audit system logs daily to detect suspicious activity as soon as possible.

9. **Perform technical audits of SCADA devices and networks and any other connected networks to identify security concerns.**

Technical audits of SCADA devices and networks are critical to ongoing security effectiveness. Many commercial and open-sourced security tools are available that allow system administrators to conduct audits of their systems/networks to identify active services, patch level, and common vulnerabilities. The use of these tools will not solve systemic problems but will eliminate the "paths of least resistance" that an attacker could exploit. Analyze identified vulnerabilities to determine their significance, and take corrective actions as appropriate. Track corrective actions, and analyze this information to identify trends. Additionally, retest systems after corrective actions have been taken to ensure that vulnerabilities were actually eliminated. Scan nonproduction environments actively to identify and address potential problems.

10. **Conduct physical security surveys and assess all remote sites connected to the SCADA network to evaluate their security.**

Any location that has a connection to the SCADA network is a target, especially unmanned or unguarded remote sites. Conduct a physical security survey and inventory access points at each facility that has a connection to the SCADA system. Identify and assess any source of information, including remote telephone/computer network/fiber-optic cables that could be tapped, radio and microwave links that are exploitable, computer terminals that could be accessed, and wireless local area network access points. Identify and eliminate single points of failure. The security of the site must be adequate to detect or prevent unauthorized access. Do not allow "live" network access points at remote, unguarded sites simply for convenience.

11. **Establish SCADA "Red Teams" to identify and evaluate possible attack scenarios.**
Establish a "Red Team" to identify potential attack scenarios and evaluate potential system vulnerabilities. Use a variety of people who can provide insight into weaknesses of the overall network, SCADA system, physical systems, and security controls. People who work on the system every day have great insight into the vulnerabilities of your SCADA network and should be consulted when identifying potential attack scenarios and possible consequences. Also, ensure that the risk from a malicious insider is fully evaluated, given that this represents one of the greatest threats to an organization. Feed information resulting from the "Red Team" evaluation into risk management processes to assess the information and establish appropriate protection strategies.

The following steps focus on management actions to establish an effective cyber security program:

12. **Clearly define cyber security roles, responsibilities, and authorities for managers, system administrators, and users.**
Organization personnel need to understand the specific expectations associated with protecting information technology resources through the definition of clear and logical roles and responsibilities. In addition, key personnel need to be given sufficient authority to carry out their assigned responsibilities. Too often, good cyber security is left up to the initiative of the individual, which usually leads to inconsistent implementations and ineffective security. Establish a cyber security organizational structure that defines roles and responsibilities and clearly identifies how cyber security issues are escalated and who is notified in an emergency.

13. **Document network architecture and identify systems that serve critical functions or contain sensitive information that require additional levels of protection.**
Develop and document robust information security architecture as part of a process to establish an effective protection strategy. It is essential that organizations design their network with security in mind and continue to have a strong understanding of their network architecture throughout its life cycle. Of particular

importance, an in-depth understanding of the functions that the systems perform and the sensitivity of the stored information is required. Without this understanding, risk cannot be properly assessed, and protection strategies may not be sufficient. Documenting the information security architecture and its components is critical to understanding the overall protection strategy and identifying single points of failure.

14. **Establish a rigorous, ongoing risk management process.**

    A thorough understanding of the risks to network computing resources from denial-of-service attacks and the vulnerability of sensitive information to compromise is essential to an effective cyber security program. Risk assessments form the technical basis of this understanding and are critical to formulating effective strategies to mitigate vulnerabilities and preserve the integrity of computing resources. Initially, perform a baseline risk analysis based on current threat assessment to use for developing a network protection strategy. Due to rapidly changing technology and the emergence of new threats on a daily basis, an ongoing risk assessment process is also needed so that routine changes can be made to the protection strategy to ensure it remains effective. Fundamental to risk management is identification of residual risk with a network protection strategy in place and acceptance of that risk by management.

15. **Establish a network protection strategy based on the principle of defense-in-depth.**

    A fundamental principle that must be part of any network protection strategy is defense-in-depth. Defense-in-depth must be considered early in the design phase of the development process and must be an integral consideration in all technical decision making associated with the network. Utilize technical and administrative controls to mitigate threats from identified risks to as great a degree as possible at all levels of the network. Single points of failure must be avoided, and cyber security defense must be layered to limit and contain the impact of any security incidents. Additionally, each layer must be protected against other systems at the same layer. For example, to protect against the inside threat, restrict users to access only those resources necessary to perform their job functions.

16. **Clearly identity cyber security requirements.**

    Organizations and companies need structured security programs with mandated requirements to establish expectations and allow personnel to be held accountable. Formalized policies and procedures are typically used to establish and institutionalize a cyber security program. A formal program is essential for establishing a consistent, standards-based approach to cyber security through an organization and eliminates sole dependence on individual initiative. Polices and procedures also inform employees of their specific cyber security responsibilities and the consequences of failing to meet those responsibilities. They also provide guidance regarding actions to be taken during a cyber security incident and promote efficient and effective actions during a time of crisis. As part of identifying cyber security requirements, include user agreements and notification and warning banners. Establish requirements to minimize the threat from malicious insiders, including the need for conducting background checks and limiting network privileges to those absolutely necessary.

17. **Establish effective configuration management processes.**

    A fundamental management process needed to maintain a secure network is configuration management. Configuration management needs to cover both hardware configurations and software configurations. Changes to hardware or software can easily introduce vulnerabilities that undermine network security. Processes are required to evaluate and control any change to ensure that the network remains secure. Configuration management begins with well-tested and documented security baselines for your various systems.

18. **Conduct routine self-assessments.**

    Robust performance evaluation processes are needed to provide organizations with feedback on the effectiveness of cyber security policy and technical implementation. A sign of a mature organization is one that is able to self-identify issues, conduct root cause analyses, and implement effective corrective actions that address individual and systemic problems. Self-assessment processes that are normally part of an effective cyber security program include routine scanning for vulnerabilities, automated auditing of the network, and self-assessments of organizational and individual performance.

19. **Establish system backups and disaster recovery plans.**

    Establish a disaster recovery plan that allows for rapid recover from any emergency (including a cyber attack). System backups are an essential part of any plan and allow rapid reconstruction of the network. Routinely exercise disaster recovery plans to ensure that they work and that personnel are familiar with them. Make appropriate changes to disaster recovery plans based on lessons learned from exercises.

20. **Senior organizational leadership should establish expectations for cyber security performance and hold individuals accountable for their performance.**

    Effective cyber security performance requires commitment and leadership from senior managers in the organization. It is essential that senior management establish an expectation for strong cyber security and communicate this to their subordinate managers throughout the organization. It is also essential that senior organizational leadership establish a structure for implementation of a cyber security program. This structure will promote consistent implementation and the ability to sustain a strong cyber security program. It is then important for individuals to be held accountable for their performance as it relates to cyber security. This includes managers, system administrators, technicians, and users/operators.

21. **Establish policies and conduct training to minimize the likelihood that organizational personnel will inadvertently disclose sensitive information regarding SCADA system design, operations, or security controls.**

Release data related to the SCADA network only on a strict, need-to-know basis, and only to persons explicitly authorized to receive such information. "Social engineering," the gathering of information about a computer or computer network via questions to naive users, is often the first step in a malicious attack on computer networks. The more information revealed about a computer or computer network, the more vulnerable the computer/network is. Never divulge data revealed to a SCADA network, including the names and contact information about the system operators/administrators, computer operating systems, and/or physical and logical locations of computers and network systems over telephones or to personnel unless they are explicitly authorized to receive such information. Any requests for information by unknown persons need to be sent to a central network security location for verification and fulfillment. People can be a weak link in an otherwise secure network. Conduct training and information awareness campaigns to ensure that personnel remain diligent in guarding sensitive network information, particularly their passwords.

REFERENCES AND RECOMMENDED READING

Brown, A. S. 2008. *SCADA vs. the hackers.* American Society of Mechanical Engineers, http://www.memagazine.org/backissues/dec02/features/scadavs/ (accessed May 10, 2008).

CRS. 2003. *Computer attack and cyber terrorism: Vulnerabilities and policy issues for Congress.* Washington, DC: Congressional Research Service.

DOE, 2001. *21 Steps to improve cyber security of SCADA networks.* Washington, DC: U.S. Department of Energy.

Ezell, B. C. 1998. *Risks of cyber attack to supervisory control and data acquisition.* Charlottesville, VA: University of Virginia.

FBI. 2000. *Threat to critical infrastructure.* Washington, DC: Federal Bureau of Investigation.

———. 2004. *Ninth annual computer crime and security survey.* Washington, DC: Computer Crime Institute and Federal Bureau of Investigations.

GAO. 2003. *Critical infrastructure protection: Challenges in securing control systems.* Washington, DC: U.S. General Accounting Office.

Gellman, B. 2002. Cyber-attacks by Al Qaeda feared: Terrorists at threshold of using Internet as tool of bloodshed, experts say. *Washington Post*, June 27, A-1.

NIPC. 2002. *National Infrastructure Protection Center report.* Washington, DC: National Infrastructure Protection Center.

Poulsen, K. 2003. *Slammer worm crashed Ohio nuke plant network.* Security Focus, http://www.securityfocus.com/news/6767 (accessed December 26, 2009).

Stamp, J., et al. 2003. *Common vulnerabilities in critical infrastructure control systems.* 2nd edition. Sandia National Laboratories.

U.S. EPA. 2005. EPA needs to determine what barriers prevent water systems from securing known SCADA vulnerabilities. In J. Harris, *Final Briefing Report.* Washington, DC: U.S. Environmental Protection Agency.

Warwalking. 2003. http://warwalking.tribe.net (accessed May 9, 2008).

Young, M. A. 2004. *SCADA Systems Security.* SANS Institute.

9

Security Techniques and Hardware

There are but two powers in the world, the sword and the mind. In the long run the sword is always beaten by the mind.

—*Napoleon Bonaparte*

Ideally, in a perfect world, all critical infrastructure sector sites/facilities would be secured in a layered fashion (a.k.a. the barrier approach). Layered security systems are vital. Using the protection-in-depth principle, requiring that an adversary defeat several protective barriers or security layers to accomplish its goal, critical infrastructure sector infrastructure can be made more secure. Protection-in-depth is a term commonly used by the military to describe security measures that reinforce one another, masking the defense mechanisms from the view of intruders, and allowing the defender time to respond to intrusion or attack.

A prime example of the use of the multibarrier approach to ensure security and safety is demonstrated by the practices of the bottled water industry. In the aftermath of 9/11 and the increased emphasis on homeland security, a shifted paradigm of national security and vulnerability awareness has emerged. Recall that in the immediate aftermath of the 9/11 tragedies, emergency responders and others responded quickly and worked to exhaustion. In addition to the emergency responders, bottled water companies responded immediately by donating several million bottles of water to the crews at the crash sites in New York, at the Pentagon, and in Pennsylvania. International Bottled Water Association (IBWA 2004) reports that "within hours of the first attack, bottled water was delivered where it mattered most; to emergency personnel on the scene who required ample water to stay hydrated as they worked to rescue victims and clean up debris" (2).

Bottled water companies continued to provide bottled water to responders and rescuers at the 9/11 sites throughout the postevent(s) process(es). These patriotic actions by the bottled water companies, however, beg the question: how do we ensure the safety and security of the bottled water provided to anyone? IBWA (2004) has the answer: using a multibarrier approach, along with other principles, will enhance the safety and security of bottled water. IBWA (2004) describes its multibarrier approach as follows:

> A multibarrier approach—Bottled water products are produced utilizing a multi-barrier approach, from source to finished product, that helps prevent possible harmful contaminants (physical, chemical or microbiological) from adulterating the finished product as well as storage, production, and transportation equipment. Measures in a multi-barrier approach may include source protection, source monitoring, reverse osmosis, distillation, filtration, ozonation or ultraviolet (UV) light. Many of the steps in a multi-barrier system may be effective in safeguarding bottled water from microbiological and other contamination. Piping in and out of plants, as well as storage silos and water tankers are also protected and maintained through sanitation procedures. In addition, bottled water products are bottled in a controlled, sanitary environment to prevent contamination during the filling operation (3).

BOTTLED WATER SECURITY VERSUS NUCLEAR POWER SECURITY

At this point the obvious question hounding most readers and burning in others is "What does bottled water security have to do with nuclear power security? The comparison is not even close—makes absolutely no sense—and is about as ludicrous a comparison as can possibly be made." Our response: Really? Consider the bottled water factory that produces, ships, and distributes one hundred thousand bottles of poisoned water to consumers. The consumers, all hundred thousand of them, either die right away, die slowly, or just vegetate until their end is reached. Consider the neighborhood nuclear power plant in your town of one hundred thousand people that is attacked by terrorists with the resulting damage leaking deadly radiation everywhere. The neighbors, all hundred thousand of them, either die right away, die slowly, or just vegetate until their end is reached.

While it is true that both drinking water and nuclear power are both important members of critical infrastructure sectors, and having made an important point (i.e., all sectors of critical infrastructure are in fact "critical") this text is about the security of nuclear power and its related ancillaries (e.g., storage and/or transportation of nuclear waste); therefore, we again shift the focus of this discussion back to the securing of nuclear power plants from terrorist attacks.

Before delving into our discussion of nuclear power plant physical security procedures, practices, and protocols, it is important to review a few key facts about nuclear power plant security, both in general and in particular.

NEI (2009) lists the following key general points:

- The nuclear energy industry is one of the few industries whose security program is regulated by the federal government (U.S. NRC).
- Nuclear plant security requirements are predicated on the need to protect the public from the possibility of exposure to radioactive releases caused by acts of sabotage, including cyber attacks.
- After the terrorist attacks of 9/11, the NRC ordered substantial security enhancements at all nuclear power plants.
- The NRC coordinates closely with other federal organizations.
- In its 2008 report to Congress on security, the NRC said, "The commission is confident that nuclear power plants . . . continue to be among the best-protected private sector facilities in the nation."

NEI (2009) also points out that key nuclear plant security measures at the 104 operating nuclear plants include

- physical barriers and illuminated detection zones
- approximately eight thousand well-trained and well-equipped armed security officers at sixty-five sites who are on duty shifts all day, every day
- surveillance and patrols of the perimeter fence
- intrusion detection aides (including several types of detection fields, closed-circuit television systems, and alarm/alert devices)
- bullet-resisting barriers to critical areas
- a dedicated contingency response force

PHYSICAL PROTECTION

According to NRC (2009), physical protection (aka physical security) consists of a variety of measures for the protection of nuclear material or facilities against sabotage, theft, and diversion. NRC's approach to physical protection is graded based on the significance of the material or facilities being protected. NRC establishes the requirements and assesses compliance with the requirements; the licensees are responsible for providing the protection. Note that NRC has a threat assessment program to determine how much physical protection is enough.

Protection of Nuclear Facilities

Nuclear facilities that require physical protection include nuclear reactors, fuel cycle facilities, and spent fuel storage and disposal facilities. Key features of the physical protection programs for these facilities include

- Protection or defense-in-depth using graded physical protection areas—Exclusion Area, Protected Area, Vital Area, and Material Access Area—barriers and controls
- Intrusion detection
- Assessment of detection alarms to distinguish between false or nuisance alarms and actual intrusions and to initiate response
- Response to intrusions
- Off-site assistance, as necessary, from local, state, and federal agencies

Protection of Nuclear Material in Transit

Transportation of spent nuclear fuel and other high-activity shipments require physical protection. Key features of physical protection for transportation include

- Use of NRC-certified, structurally rugged shipment overpacks and canisters. Fuel within canisters is dense and in solid form, not readily dispersible as respirable particles.
- Advance planning and coordination with local law enforcement along approved routes
- Protection of information about schedules
- Regular communication between transports and control centers
- Armed escorts within heavily populated areas
- Vehicle immobility measures to protect against movement of a hijacked shipment before response forces arrive

THREAT ASSESSMENT

To determine how much physical protection is enough, the NRC monitors intelligence information to keep abreast of foreign and domestic events and remains aware of the capabilities of potential adversaries. The NRC uses this information, and other sources, to determine the physical protection requirements of their regulations and to establish and maintain design basis threats (DBTs). Nuclear power plants and selected fuel cycle facilities must be defended against the DBTs. The DBTs may be found in NRC regulation 10 CFR 73.1. Pertinent physical security definitions listed in the following section are found in section 73.2.

Definitions

armed escort—an armed person, not necessarily uniformed, whose primary duty is to accompany shipments of special nuclear material for the protection of such shipments against theft or radiological sabotage.

armed response personnel—persons, not necessarily uniformed, whose primary duty in the event of attempted theft of special nuclear material or radiological sabotage is to respond, armed and equipped, to prevent or delay such actions.

authorized individual—any individual, including an employee, a student, a consultant, or an agent of a licensee who has been designated in writing by a licensee to have responsibility for surveillance of or control over special nuclear material or to have unescorted access to areas where special nuclear material is used or stored.

background check—at a minimum, a Federal Bureau of Investigation (FBI) criminal history records check (including verification of identity based on fingerprinting), employment history, education, and personal references. Individuals engaged in activities subject to regulation by the commission, applicants for license to engage in commission-regulated activities, and individuals who have notified the commission in writing of an intent to file an application for licensing, certification, permitting, or approval of a product or activity subject to regulation by the commission are required under section 73.57 to conduct fingerprinting and criminal history records checks before granting access to Safeguards Information. A background check must be sufficient to support the trustworthiness and reliability determination so that the person performing the check and the commission have assurance that granting individuals access to Safeguards Information does not constitute an unreasonable risk to the public health and safety or the common defense and security.

bullet-resistant—protection against complete penetration, passage of fragments of projectiles, and spalling (fragmentation) of the protective material that could cause injury to a person standing directly behind the bullet-resisting barrier.

contiguous sites—licensee-controlled locations, deemed by the commission to be in close enough proximity to each other, that the special nuclear material must be considered in the aggregate for the purpose of physical protection.

continuous visual surveillance—unobstructed view at all times of a shipment of special nuclear material, and of all access to a temporary storage area or cargo compartment containing the shipment.

controlled access area—any temporarily or permanently established area that is clearly demarcated, access to which is controlled, and that affords isolation of the material or persons within it.

deceit—methods used to attempt to gain unauthorized access, introduce unauthorized materials, or remove strategic special nuclear materials, where the attempt involves falsification to present the appearance of authorized access.

force—violent methods used by an adversary to attempt to steal strategic special nuclear material or to sabotage a nuclear facility, or violent methods used by response personnel to protect against such adversary actions.

formula quantity—strategic special nuclear material in any combination in a quantity of five thousand grams or more computed by the formula, grams = (grams contained U-235) + 2. grams U-233 + grams plutonium). This class of material is sometimes referred to as a Category 1 quantity of material.

guard—a uniformed individual armed with a firearm whose primary duty is the protection of special nuclear material against theft, the protection of a plant against radiological sabotage, or both.

incendiary device—any self-contained device intended to create an intense fire that can damage normally flame-resistant or flame-retardant materials.

individual authorized access to Safeguards Information—an individual authorized to have access to and handle such information pursuant to the requirements of sections 73.21 and 73.22.

intrusion alarm—a tamper indicating electrical, electromechanical, electro-optical, electronic, or similar device that will detect intrusion by an individual into a building, protected area, vital area, or material access area and alert guards or watchmen by means of actuated visible and audible signals.

isolation zone—any area adjacent to a physical barrier, clear of all objects that could conceal or shield an individual.

lock—in the case of vaults or vault-type rooms, a three-position, manipulation-resistant, dial-type, built-in combination lock or combination padlock; in the case of fences, walls, and buildings, an integral door lock or padlock that provides protection equivalent to a six-tumbler cylinder lock. Lock in the case of a vault or vault-type room also means any manipulation-resistant, electromechanical device that provides the same function as a built-in combination lock or combination padlock, which can be operated remotely or by the reading or insertion of information, which can be uniquely characterized, and which allows operation of the device. Locked means protected by an operable lock.

material access area—any location that contains special nuclear material, within a vault or a building, the roof, walls, and floor of which each constitute a physical barrier.

movement control center—an operations center that is remote from transport activity and that maintains periodic position information on the movement of strategic special nuclear material, receives reports of attempted attacks or thefts, provides a means for reporting these and other problems to appropriate agencies, and can request and coordinate appropriate aid.

need to know—a determination by a person having responsibility for protecting Safeguards Information (including Safeguards Information designated as Safeguards Information—Modified Handling) that a proposed recipient's access to Safeguards Information is necessary in the performance of official, contractual, licensee, applicant, or certificate holder employment. In an adjudication, "need to know" means a determination by the originator of the information that the information is necessary to enable the proposed recipient to proffer and/or adjudicate a specific contention in that proceeding, and the proposed recipient of the specific Safeguards Information possesses demonstrable knowledge, skill, training, or education to effectively utilize the specific Safeguards Information in the proceeding. Where the information is in the possession of the originator and the NRC staff (dual possession), whether in its original form or incorporated into another document or other matter by the recipient, the NRC staff makes the determination. In the event of a dispute regarding the "need to know" determination, the presiding officer of the proceeding will make the "need to know" determination.

physical barrier—(1) fences constructed of no. 11 American wire gauge, or heavier wire fabric, topped by three strands or more of barbed wire or similar material on brackets angled inward or outward between thirty and forty-five degrees from the vertical, with an overall height of not less than eight feet, including the barbed topping; (2) building walls, ceilings, and floors constructed of stone, brick, cinder block, concrete, steel, or comparable materials (openings in which are secured by grates, doors, or covers of construction and fastening of sufficient strength such that the integrity of the wall is not lessened by any opening), or walls of similar construction, not part of a building, provided with a barbed topping described in paragraph (1) of this definition of a height of not less than eight feet; or (3) any other physical obstruction constructed in a manner and of materials suitable for the purpose of which the obstruction is intended.

protected area—an area encompassed by physical barriers and to which access is controlled.

radiological sabotage—any deliberate act directed against a plant or transport in which an activity licensed pursuant to the regulations in this chapter is conducted, or against a component of such a plant or transport that could directly or indirectly endanger the public health and safety by exposure to radiation.

Safeguards Information—information not classified as National Security Information or Restricted Data that specifically identifies a licensee's or applicant's detailed control and accounting procedures for the physical protection of special nuclear material in quantities determined by the commission through order or regulation to be significant to the public health and safety or the common defense and security; detailed security measures (including security plans, procedures, and equipment) for the physical protection of source, by-product, or special nuclear material in quantities determined by the commission through order or regulation to be significant to the public health and safety or the common defense and security; security measures for the physical protection of and location of certain plant equipment vital to the safety of production or utilization facilities; and any other information within the scope of section 147 of the Atomic Energy Act of 1954, as amended, the unauthorized disclosure of which, as determined by the commission through order or regulation, could reasonably be expected to have a significant adverse effect on the health and safety of the public or the common defense and security by significantly increasing the likelihood of sabotage or theft or diversion of source, by-product, or special nuclear material.

Security Management—persons responsible at the policy and general management level.

Security Storage Container—includes any of the following repositories: (1) For storage in a building located within a protected or controlled access area, a steel filing cabinet equipped with a steel locking bar and a three-position, changeable combination, GSA approved padlock; (2) A security filing cabinet that bears a Test Certification Label on the side of the locking drawer or interior plate and is marked "General Services Administration Approved Security Container" on the exterior of the top drawer or door; (3) A bank safety deposit box; (4) Other repositories which in the judgment of the NRC would provide comparable physical protection.

security supervision—persons, not necessarily uniformed or armed, whose primary duties are supervision and direction of security at the day-to-day operating level.

special nuclear material of low strategic significance—(1) Less than an amount of special nuclear material of moderate strategic significance as defined in paragraph (1) of the definition of strategic nuclear material of moderate strategic significance in this section, but more than fifteen grams of uranium-235 (continued in uranium enriched to 20 percent or more in U-235 isotope) or fifteen grams of uranium-233 or fifteen grams of plutonium or the combination of fifteen grams when computed by the equation grams = (grams contained U-235) + (grams plutonium) + (grams U-233); (2) Less than ten thousand grams but more than one thousand grams of uranium-235 (continued in uranium enriched to 10 percent or more but less than 20 percent in the U-235 isotope); or (3) 10,000 grams or more of uranium-235 (contained in uranium enriched above natural but less than 10 percent in the U-235 isotope). This class of material is sometimes referred to as a Category III quantity of material.

special nuclear material of moderated strategic significance—(1) Less than a formula quantity of strategic special nuclear material but more than one thousand grams of uranium-235 (contained in uranium enriched to 20 percent or more in the U-235 isotope) or more than five hundred grams of uraninum-233, or plutonium, or in a combined quantity of more than one thousand grams when computed by the equation grams = (grams continued U-235) + 2 (grams U-233 + grams plutonium); (2) Ten thousand grams or more of uranium-235 (contained in uranium enriched to 10 percent or more but less than 20 percent in the U-235 isotope). This class of material is sometimes referred to as a Category II quantity of material.

stealth—methods used to attempt to gain unauthorized access, introduce unauthorized materials, or remove strategic special nuclear material, where the fact of such attempt is concealed or an attempt is made to conceal it.

strategic special nuclear material—uranium-235 (contained in uranium enriched to 20 percent or more in the U-235 isotope), uranium-233, or plutonium.

Tactical Response team—the primary response force for each shift and that can be identified by a distinctive item of uniform, armed with specified weapons, and whose other duties permit immediate response.

transport—any land, sea, or air conveyance or modules for these conveyances, such as rail cars or standardized cargo containers.

trustworthiness and reliability—characteristics of an individual considered dependable in judgment, character, and performance, such that disclosure of Safeguards Information (including Safeguards Information designated as Safeguards Information—Modified Handling) to that individual does not constitute an unreasonable risk to the public health and safety or common defense and security. A determination of trustworthiness and reliability for this purpose is based upon a background check.

undergoing processing—performing active operations on material such as chemical transformation, physical transformation, or transit between such operations, to be differentiated form storage or packaging for shipment.

vault—a windowless enclosure with walls, floor, roof, and door(s) designed and constructed to delay penetration in case of forced entry.

vault-type room—a room with one or more doors, all capable of being locked, protected by an intrusion alarm that creates an alarm upon the entry of a person anywhere into the room and upon exit from the room or upon movement of an individual within the room.

vital area—any area that contains vital equipment.

vital equipment—any equipment, system, device, or material, the failure, destruction, or release of which could directly or indirectly endanger the public health and safety by exposure to radiation. Equipment or systems that would be required to function to protect public health and safety following such failure, destruction, or release are also considered to be vital.

watchman—an individual, not necessarily uniformed or armed with a firearm, who provides protection for a plant and the special nuclear material therein in the course of performing other duties.

PHYSICAL PROTECTION AREAS

In nuclear power sector infrastructure security, protection-in-depth is used to describe a layered security approach. A protection-in-depth strategy uses several forms of security techniques and/or devices against an intruder and does not rely on one single defensive mechanism to protect infrastructure. By implementing multiple layers of security, a hole or flaw in one layer is covered by the other layers. An intruder will have to intrude through each layer without being detected in the process—the layered approach implies that no matter how an intruder attempts to accomplish his goal, he will encounter effective elements of the physical protection system.

As mentioned, NRC requires licensees to take a graded approach to physical protection through use of specifically defined areas with increasing levels of security. These areas include

- Exclusion area
- Protected area
- Vital access area
- Material access area

Exclusion Area

The exclusion area is that area in which the licensee has the authority to determine all activities, including exclusion or removal of personnel and property from the area.

- Although the licensee may have fences and guard posts to limit access to an exclusion area, these are not required as long as the licensee controls the area.
- This area may be traversed by a highway, railroad, or waterway, provided arrangements are made to control traffic on the highway, railroad, or water in case of emergency.
- Homes are not normally permitted within the exclusion area.

Protected Area

The protected area is an area within the exclusion area encompassed by physical barriers, such as one or more chain link fences.

- Access to a protected area is controlled at gates and entryways and attempts to access under, over, or through the barrier are detected by the perimeter intrusion detection.
- Authorization for unescorted access within the protected area is based on criminal history and other background and behavioral checks, such as fitness-for-duty programs.

Vital Areas

Vital areas are located within protected areas and have additional barriers and alarms to protect vital equipment.

- Additional authorization is required for unescorted access to vital areas.
- Entry is through card-reader or security-guard-controlled doors.

Material Access Areas

Material access areas are similar to vital areas but control access to forms of special nuclear material for which protection against theft and diversion is required. The physical protection for these areas is similar to that of vital areas with the additional conditions that

- No one is allowed to be alone in a material access area (two-person rule).
- In addition to card-reader or security-guard-controlled doors, volumetric intrusion detection systems are used when the area is unoccupied.

INTRUSION DETECTION

An intrusion detection system is intended to notify the site's security force of a potential intruder. Licensee detection capability addresses the following adversary intrusion tactics:

- stealth
- deceit
- force

Perimeter intrusion detection systems include:

- Volumetric systems, such as microwave and some electric field systems
- Volumetric systems detect disturbances of a volume of space.
- For a microwave system with separate transmitters and receivers, this space is between the transmitter and the receiver.
- For an electric field sensor, the space parallels the sensor and generator wires.
- The effective detection volume must be sufficiently high and wide to prevent jumping across without detection and sufficiently low to detect someone attempting to crawl beneath it.
- Planar systems, such as fence disturbance and infrared systems, detect disturbances of the geometric plane containing the system elements.
- Security officers

Interior intrusion detection systems include:

- Door-mounted balanced magnetic switches
- Interior volumetric systems
- Ultrasound
- Infrared

- Motion detection closed circuit television

ASSESSMENT

Intrusion detection systems can incorrectly alarm because of environmental or other factors; therefore, the licensee must assess whether an alarm is genuine or not. The assessment also provides information to the security force about the number, location, and weapons of the intruders.

Intrusion alarm assessment can be performed in a number of ways, including use of

- Closed circuit television (CCTV) cameras to observe the zone of the alarm
- CCTV performance can be enhanced by use of video capture systems to allow instant playback of the scene at the time that alarm was tripped.
- Security officers in guard towers
- Security officers proceeding to the alarm zone

Intrusion alarm assessment performance should be adequate to correctly identify actual intrusions despite adversary use of such techniques as

- speed of individuals through the isolation zone
- camouflage of persons in the isolation zone
- shadows or other lighting conditions in the isolation zone
- alarm stacking, that is, causing a large number of alarms over a short period of time
- normal environmental factors affecting visibility

ARMED RESPONSE

To protect public health and safety and the common defense and security, the licensee is responsible for defending nuclear material or a nuclear facility against an attack, using armed responses. This may include the use of deadly force if necessary.

Response Capabilities

- Fixed sites: Depending upon the significance of the material or facilities being protected, armed response to an unauthorized intrusion into a protected area and attack on a nuclear facility could be by the Licensee's on-site armed security force, with the local law enforcement authority (LLEA) and FBI arriving later
- LLEA
- Transportation: Armed respond to an attack on a nuclear transport would be by the
- Licensee's armed escorts in the case of a shipment of Category 1 material (strategically significant SNM), with FBI and LLEA arriving later

- Licensee's armed escorts and LLEA in the case of a shipment of spent fuel or other high-activity shipments

Response Preparation

- Response elements must be appropriately armed and in sufficient number to counter the potential threat.
- A licensee's armed security officers undergo significant training and qualification in, among other things,
  - physical fitness
  - weapons safety
  - nuclear and radiation
  - weapons proficiency
  - use of force
  - tactical movements
  - response struggles
- Because of redundancy of safety equipment at nuclear facilities, the plant may elect a protection strategy that protects only a subset of vital equipment. The subset is determined in advance to be sufficient to permit safe shutdown of the reactor or facility. The subset to protect may vary, depending upon where the adversary enters the protected area.
- Interaction with LLEA includes advance planning and coordination, including familiarity tours.

SECURITY HARDWARE DEVICES*

In the following sections, various security hardware and/or devices are described. These devices serve the main purpose of providing security against physical and/or digital intrusion. That is, they are designed to delay and deny intrusion and are normally coupled with detection and assessment technology. Possible additional security measures, based on the vulnerability assessment that may be recommended (covered in this text) include the following (NAERC 2002):

- Electronic security
- Closing nonessential perimeter and internal portals
- Physical barriers such as bollards or Jersey walls
- Fencing
- Lighting

---

* It is important to point out that even though the following U.S. EPA security asset and device recommendations were first made for the water/wastewater critical infrastructure, these recommendations are applicable to all other critical infrastructure sectors, including the nuclear power sector.

- Security surveys
- Vulnerability assessments
- Availability of security resources
- General personnel and security officer training
- Law enforcement liaison
- Ensuring availability of essential spare parts (wire, pipe, valves, transformers, etc.) for critical facilities

Keep in mind, however, and as mentioned previously, no matter the type of security device or system employed, nuclear power sector systems cannot be made immune to all possible intrusions or attacks—nothing can be made absolutely secure. Whenever a plant or facility safety/security manager tells us that he or she has secured the site 100 percent, we are reminded of Schneier's (2000) view of security: "You can't defend. You can't prevent. The only thing you can do is detect and respond." Simply, when it comes to making "anything" absolutely secure from intrusion or attack, there is inherently, or otherwise, no silver bullet.

U.S. EPA (2005) groups the infrastructure security devices or products described below into four general categories:

- Physical asset monitoring and control devices
- Cyber protection devices
- Communication/integration
- Environmental monitoring devices

Physical Asset Monitoring and Control Devices

Aboveground, Outdoor Equipment Enclosures

Experience demonstrates that many different system components can be and are often installed outdoors and aboveground. Examples of these types of components could include

- Backflow prevention devices
- Air release and control valves
- Pressure vacuum breakers
- Oil and gas pumps and motors
- Cooling water storage and feed equipment
- Meters
- Sampling equipment
- Instrumentation
- Electrical substations

One of the most effective security measures for protecting aboveground equipment, where feasible, is to place it inside a building or exterior fenced structure. When/where this is not possible, enclosing the equipment or parts of the equipment using some sort of commercial or homemade add-on structure may help to prevent tampering with the equipment. These types of add-on structures or enclosures, which are designed to protect people and animals from electrocution and to protect equipment both from the elements and from unauthorized access or tampering, typically consist of a box-like fenced structure that is placed over or around the entire component, or over/around critical parts of the component (i.e., valves, etc.), and is then secured to delay or prevent intruders from tampering with the equipment. The enclosures are typically locked or otherwise anchored to a solid foundation, which makes it difficult for unauthorized personnel to remove the enclosure and access the equipment.

Standardized aboveground enclosures are available in a wide variety of materials, sizes, and configurations. Many options and security features are also available for each type of enclosure, and this allows system operators the flexibility to customize an enclosure for a specific application and/or price range. In addition, most manufacturers can custom-design enclosures if standard, off-the-shelf enclosures do not meet a user's needs.

Many of these enclosures are designed to meet certain standards. For example, the American Society of Sanitary Engineers (ASSE) has developed Standard #1060, *Performance Requirements for Outdoor Enclosures for Backflow Prevention Assemblies*. If an enclosure will be used to house backflow preventer, this standard specifies the acceptable construction materials for the enclosure, as well as the performance requirements that the enclosure should meet, including specifications for freeze protection, drainage, air inlets, access for maintenance, and hinge requirements. ASSE #1060 also states that the enclosure should be lockable to enhance security.

Electrical substation and electrical equipment enclosures must meet the requirements and recommendations of various OHSA, National Fire Protection Association (NFPA), National Electrical Codes (NEC), Institute of Electrical and Electronic Engineers (IEEE), and local code requirements.

Equipment enclosures can generally be categorized into one of four main configurations, which include

- One-piece drop-over enclosures
- Hinged- or removable-top enclosures
- Sectional enclosures
- Shelters with access locks

All enclosures, including those with integral floors, must be secured to a foundation to prevent them from being moved or removed. Unanchored or poorly anchored enclosures may be blown off the equipment being protected or may be defeated by intruders. In either case, this may result in the equipment beneath the enclosure becoming exposed and damaged. Therefore, ensuring that the enclosure is securely anchored will increase the security of the protected equipment.

The three basic types of foundations that can be used to anchor the aboveground equipment enclosure are concrete footers, concrete slabs-on-grade, or manufactured fiberglass pads. The most common types of foundations utilized for equipment enclosures are standard or slab-on-grade footers; however, local climate and soil conditions may dictate whether either of these types of foundations can be used. These foundations can be either precast or poured in place at the installation site. Once the foundation is installed and properly cured, the equipment enclosure is bolted or anchored to the foundation to secure it in place.

An alternative foundation, specifically for use with smaller Hot Box enclosures, is a manufactured fiberglass pad known as the Glass Pad. The Glass Pad has the center cut out so that it can be dropped directly over the piece of equipment being enclosed. Once the pad is set level on the ground, it is backfilled over a two-inch flange located around its base. The enclosure is then placed on top of the foundation, and is locked in place with either a staple- or a slotted-anchor, depending on the enclosure configuration.

One of the primary attributes of a security enclosure is its strength and resistance to breaking and penetration. Accordingly, the materials from which the enclosure is constructed will be important in determining the strength of the enclosure and thus its usefulness for security applications. Enclosures are typically manufactured for either fiberglass or aluminum. With the exception of the one-piece, drop-over enclosure, which is typically fabricated from fiberglass, each configuration described above can be constructed from either material. In addition, enclosures can be custom-manufactured from polyurethane, galvanized steel, or stainless steel. Galvanized or stainless steel is often offered as an exterior layer, or "skin," for an aluminum enclosure. Although they are typically utilized in underground applications, precast concrete structures can also be used as aboveground equipment enclosures. However, precast structures are much heavier and more difficult to maneuver than are their fiberglass and aluminum counterparts. Concrete is also brittle, and that can be a security concern; however, products can be applied to concrete structures to add strength and minimize security risks (e.g., epoxy coating). Because precast concrete structures can be purchased from any concrete producers, this document does not identify specific vendors for these types of products.

In addition to the construction materials, enclosure walls can be configured or reinforced to give them added strength. Adding insulation is one option that can strengthen the structural characteristics of an enclosure; however, some manufacturers offer additional features to add strength to exterior walls. For example, while most enclosures are fabricated with a flat wall construction, some vendors manufacture fiberglass shelters with ribbed exterior walls. These ribs increase the structural integrity of the wall and allow the fabrication of standard shelters up to twenty feet in length. Another vendor has developed a proprietary process that uses a series of integrated fiberglass beams that are placed throughout a foam inner core to tie together the interior and exterior walls and roof. Yet another vendor constructs aluminum enclosures with horizontal and vertical redwood beams for structural support.

Other security features that can be implemented on aboveground, outdoor equipment enclosures include locks, mounting brackets, tamper-resistant doors, and exterior lighting.

Active Security Barriers (Crash Barriers)

Active security barriers (also known as crash barriers) are large structures that are placed in roadways at entrance and exit points to protected facilities by controlling vehicle access to these areas. These barriers are placed perpendicular to traffic to block the roadway so that the only way that traffic can pass the barrier is for the barrier to be moved out of the roadway. These types of barriers are typically constructed from sturdy materials, such as concrete or steel, such that vehicles cannot penetrate through them. They are also designed at a certain height off the roadway so that vehicles cannot go over them.

The key difference between active security barriers, which include wedges, crash beams, gates, retractable bollards, and portable barricades, and passive security barriers, which include nonmovable bollards, jersey barriers, and planters, is that active security barriers are designed so that they can be raised and lowered or moved out of the roadway easily to allow authorized vehicles to pass them. Many of these types of barriers are designed so that they can be opened and closed automatically (i.e., mechanized gates, hydraulic wedge barriers), while others are easy to open and close manually (swing crash beams, manual gates). In contrast to active barriers, passive barriers are permanent, nonmovable barriers, and thus they are typically used to protect the perimeter of a protected facility, such as sidewalks and other areas that do not require vehicular traffic to pass them. Several of the major types of active security barriers, such as wedge barriers, crash beams, gates, bollards, and portable/removable barricades, are described below.

Wedge barriers are plated, rectangular steel buttresses approximately two to three feet high that can be raised and lowered from the roadway. When they are in the open position, they are flush with the roadway, and vehicles can pass over them. However, when they are in the closed (armed) position, they project up from the road at a forty-five-degree angle, with the upper end pointing toward the oncoming vehicle and the base of the barrier away from the vehicle. Generally, wedge barriers are constructed from heavy-gauge steel, or concrete that contains an impact-dampening iron rebar core that is strong and resistant to breaking or cracking, thereby allowing them to withstand the impact from a vehicle attempting to crash through them. In addition, both of these materials help to transfer the energy of the impact over the barrier's entire volume, thus helping to prevent the barrier from being sheared off its base. In addition, because the barrier is angled away from traffic, the force of any vehicle impacting the barrier is distributed over the entire surface of the barrier and is not concentrated at the base, which helps prevent the barrier from breaking off at the base. Finally, the angle of the barrier helps hang up any vehicles attempting to drive over it.

Wedge barriers can be fixed or portable. Fixed wedge barriers can be mounted on the surface of the roadway ("surface-mounted wedges") or in a shallow mound in the road's surface, or they can be installed completely below the road surface. Surface-mounted wedge barricades operate by rising from a flat position on the surface of the roadway, while shallow-mount wedge barriers rise from their resting position just below the road surface. In contrast, below-surface wedge barriers operate by rising from beneath the road surface. Both the shallow-mounted and surface-mounted barriers require little or no excavation and thus do not interfere with buried utilities. All three barrier mounting types project above the road surface and block traffic when they are raised into the armed position. Once they are disarmed and lowered, they are flush with the road, thereby allowing traffic to pass. Portable wedge barriers are moved into place on wheels that are removed after the barrier has been set into place.

Installing rising wedge barriers requires preparation of the road surface. Installing surface-mounted wedges does not require that the road be excavated; however, the road surface must be intact and strong enough to allow the bolts anchoring the wedge to the road surface to attach properly. Shallow-mount and below-surface wedge barricades require excavation of a pit that is large enough to accommodate the wedge structure, as well as any arming/disarming mechanisms. Generally, the bottom of the excavation pit is lined with gravel to allow for drainage. Areas not sheltered from rain or surface runoff can install a gravity drain or self-priming pump. Table 9.1 lists the pros and cons of wedge barriers.

Crash beam barriers consist of aluminum beams that can be opened or closed across the roadway. While there are several crash beam designs, every crash beam system consists of an aluminum beam that is supported on each side by a solid footing or buttress, which is typically constructed from concrete, steel, or some other strong material. Beams typically contain an interior steel cable (typically at least one inch in diameter) to give the beam added strength and rigidity. The beam is connected by a heavy duty hinge or other mechanism to one of the footings so that it can swing or rotate out of the roadway when it is open and can swing back across the road when it

Table 9.1. Pros and Cons of Wedge Barriers

Pros	Cons
Can be surface mounted or completely installed below the roadway surface.	Installations below the surface of the roadway will require construction that may interfere with buried utilities.
Wedge barriers have a quick response time (normally 3.5–10.5 seconds, but barrier can be 1–3 seconds in emergency situations. Because emergency activation of the barrier causes more wear and tear on the system than does normal activation, it is recommended for use only in true emergency situations.	Regular maintenance is needed to keep wedge fully operational.
Surface or shallow-mount wedge barricades can be utilized in locations with a high water table and/or corrosive soils.	Improper use of the system may result in authorized vehicles being hung up by the barrier and damaged. Guards must be trained to use the system properly to ensure that this does not happen. Safety technologies may also be installed to reduce the risk of the wedge activating under an authorized vehicle.
All three wedge barrier designs have a high crash rating, thereby allowing them to be employed for higher security applications.	
These types of barrier are extremely visible, which may deter potential intruders.	

Source: U.S. Environmental Protection Agency, *Water and Wastewater Security Product Guide*, http://cfpub.epa.gov/safewater/watersecurity/guide (accessed April 4, 2009).

is in the closed (armed) position, blocking the road and inhibiting access by unauthorized vehicles. The nonhinged end of the beam can be locked into its footing, thus providing anchoring for the beam on both sides of the road and increasing the beam's resistance to any vehicles attempting to penetrate through it. In addition, if the crash beam is hit by a vehicle, the aluminum beam transfers the impact energy to the interior cable, which in turn transfers the impact energy through the footings and into their foundation, thereby minimizing the chance that the impact will snap the beam and allow the intruding vehicle to pass through.

Crash beam barriers can employ drop-arm, cantilever, or swing beam designs. Drop-arm crash beams operate by raising and lowering the beam vertically across the road. Cantilever crash beams are projecting structures that are opened and closed by extending the beam from the hinge buttress to the receiving buttress, located on the opposite side of the road. In the swing beam design, the beam is hinged to the buttress such that it swings horizontally across the road. Generally, swing beam and cantilever designs are used at locations where a vertical lift beam is impractical. For example, the swing beam or cantilever designs are utilized at entrances and exits with overhangs, trees, or buildings that would physically block the operation of the drop-arm beam design.

Table 9.2. Pros and Cons of Crash Beams

Pros	Cons
Requires little maintenance while providing long-term durability.	Crash beams have a slower response time (normally 9.5–15.3 seconds, but can be reduced to 7–10 seconds in emergency situations) than do other types of active security barriers, such as wedge barriers. Because emergency activation of the barrier causes more wear and tear on the system than does normal activation, it is recommended for use only in true emergency situations.
No excavation is required in the roadway itself to install crash beams.	All three crash beam designs possess a low crash rating relative to other types of barriers, such as wedge barriers, and thus they typically are used for lower-security applications.
	Certain crash barriers may not be visible to oncoming traffic and therefore may require additional lighting and/or other warning markings to reduce the potential for traffic to accidentally run into the beam.

Source: U.S. Environmental Protection Agency, *2005 Water and Wastewater Security Product Guide*, http://.cfpub.epa .gov/safewater/watersecurity/guide (accessed April 4, 2009).

Installing any of these crash beam barriers involves the excavation of a pit approximately forty-eight inches deep for both the hinge and the receiver footings. Due to the depth of excavation, the site should be inspected for underground utilities before digging begins. Table 9.2 lists the pros and cons of crash beams.

In contrast to wedge barriers and crash beams, which are typically installed separately from a fence line, *gates* are often integrated units of a perimeter fence or wall around a facility. Gates are basically movable pieces of fencing that can be opened and closed across a road. When the gate is in the closed (armed) position, the leaves of the gate lock into steel buttresses that are embedded in concrete foundation located on both sides of the roadway, thereby blocking access to the roadway. Generally, gate barricades are constructed from a combination of heavy-gauge steel and aluminum that can absorb an impact from vehicles attempting to ram through them. Any remaining impact energy not absorbed by the gate material is transferred to the steel buttresses and their concrete foundation.

Gates can utilize a cantilever, linear, or swing design. Cantilever gates are projecting structures that operate by extending the gate from the hinge footing across the roadway to the receiver footing. A linear gate is designed to slide across the road on tracks via a rack-and-pinion drive mechanism. Swing gates are hinged so that they can swing horizontally across the road.

Table 9.3. Pros and Cons of Gates

Pros	Cons
All three gate designs possess an intermediate crash rating, thereby allowing them to be utilized for medium- to higher-security applications.	Gates have a slower response time (normally 10-15 seconds, but can be reduced to 7-10 seconds in emergency situations) than do other types of active security barriers, such as wedge barriers. Because emergency activation of the barrier causes more wear and tear on the system than does normal activation, it is recommended for use only in true emergency situations.
Requires very little maintenance.	
Can be tailored to blend in with perimeter fencing.	
Gate construction requires no roadway excavation.	
Cantilever gates are useful for roads with high crowns or drainage gutters.	
These types of barriers are extremely visible, which may deter intruders.	
Gates can also be used to control pedestrian traffic.	

Source: U.S. Environmental Protection Agency, *Water and Wastewater Security Product Guide*, http://.cfpub.epa.gov/safewater/watersecurity/guide (accessed April 4, 2009).

Installation of the cantilever, linear, or swing gate designs described above involve the excavation of a pit approximately forty-eight inches deep for both the hinge and receiver footings to which the gates are attached. Due to the depth of excavation, the site should be inspected for underground utilities before digging begins. Table 9.3 lists the pros and cons of gates.

Bollards are vertical barriers at least three feet tall and one to two feet in diameter that are typically set four to five feet apart from each other so that they block vehicles from passing between them. Bollards can be either fixed in place, removable, or retractable. Fixed and removable bollards are passive barriers that are typically used along building perimeters or on sidewalks to prevent vehicles from driving on them while allowing pedestrians to pass them. In contrast to passive bollards, retractable bollards are active security barriers that can easily be raised and lowered to allow vehicles to pass between them. Thus, they can be used in driveways or on roads to control vehicular access. When the bollards are raised, they protect above the road surface and block the roadway; when they are lowered, they sit flush with the road surface and thus allow traffic to pass over them. Retractable bollards are typically constructed from steel or other materials that have a low weight-to-volume ratio so that they require low power to raise and lower. Steel is also more resistant to breaking than is a more brittle material, such as concrete, and is better able to withstand direct vehicular impact without breaking apart.

Retractable bollards are installed in a trench dug across a roadway—typically at an entrance or gate. Installing retractable bollards requires preparing the road surface. Depending on the vendor, bollards can be installed either in a continuous slab of concrete or in individual excavations with concrete poured in place. The required excavation for a bollard is typically slightly wider and slightly deeper than the bollard height when extended aboveground. The bottom of the excavation is typically lined with gravel to allow drainage. The bollards are then connected to a control panel that controls the raising and lowering of the bollards. Installation typically requires mechanical, electrical, and concrete work; if utility personnel with these skills are available, then the utility can install the bollards themselves. Table 9.4 lists the pros and cons of retractable bollards.

Portable/removable barriers, which can include removable crash beams and wedge barriers, are mobile obstacles that can be moved in and out of position on a roadway. For example, a crash beam may be completely removed and stored off-site when it is not needed. An additional example would be wedge barriers that are equipped with wheels that can be removed after the barricade is towed into place.

When portable barricades are needed, they can be moved into position rapidly. To provide them with added strength and stability, they are typically anchored to buttress boxes that are located on either side of the road. These buttress boxes, which may or may not be permanent, are usually filled with sand, water, cement, gravel, or concrete to make them heavy and aid in stabilizing the portable barrier. In addition, these buttresses can help dissipate any impact energy from vehicles crashing into the barrier itself.

Because these barriers are not anchored into the roadway, they do not require excavation or other related construction for installation. In contrast, they can be assembled

Table 9.4. Pros and Cons of Retractable Bollards

Pros	Cons
Bollards have a quick response time (normally three to ten seconds, but can be reduced to one to three seconds in emergency situations).	Bollard installations will require construction below the surface of the roadway, which may interfere with buried utilities.
Bollards have an intermediate crash rating, which allows them to be utilized for medium- to higher-security applications.	Some maintenance is needed to ensure that the barrier is free to move up and down.
	The distance between bollards must be decreased (i.e., more bollards must be installed along the same perimeter) to make these systems effective against small vehicles (i.e., motorcycles).

Source: U.S. Environmental Protection Agency, *Water and Wastewater Security Product Guide*, http://cfpub.epa.gov/safewater/watersecurity/guide (accessed April 4, 2009).

and made operational in a short period of time. The primary shortcoming to this type of design is that these barriers may move if they are hit by vehicles. Therefore, it is important to carefully assess the placement and anchoring of these types of barriers to ensure that they can withstand the types of impacts that may be anticipated at that location. Table 9.5 lists the pros and cons of portable/removable barricades.

Because the primary threat to active security barriers is that vehicles will attempt to crash through them, their most important attributes are their size, strength, and crash resistance. Other important features for an active security barrier are the mechanisms by which the barrier is raised and lowered to allow authorized vehicle entry and other factors, such as weather resistance and safety features.

In addition to the previously listed physical barriers, design barriers can be instituted as well. One easy and inexpensive design feature is parking lot orientation. There are many choices available for parking lot design, and most people don't put much thought into it when they park their cars. A parking lot with the rows running perpendicular to the building, however, could provide a terrorist with a runway to attempt to crash into a building or physical barrier. Simply running the rows of cars parallel with the building structure will block the path needed for a car to build momentum for an effective crash.

Alarms

An *alarm system* is a type of electronic monitoring system that is used to detect and respond to specific types of events—such as unauthorized access to an asset or a possible fire. In nuclear power plant operational systems, alarms are also used to alert operators when process operating or monitoring conditions go out of preset parameters (i.e., process alarms). These types of alarms are primarily integrated with process

Table 9.5. Pros and Cons of Portable/Removable Barricades

Pros	Cons
Installing portable barricades requires no foundation or roadway excavation.	Portable barriers may move slightly when hit by a vehicle, resulting in a lower crash resistance.
Can be moved in and out of position in a short period of time.	Portable barricades typically require 7.75 to 16.25 seconds to move into place, and thus they are considered to have a medium response time when compared with other active barriers.
Wedge barriers equipped with wheels can be easily towed into place.	
Minimal maintenance is needed to keep barriers fully operational.	

Source: U.S. Environmental Protection Agency, *Water and Wastewater Security Product Guide*, http://cfpub.epa.gov/safewater/watersecurity/guide (accessed April 4, 2009).

monitoring and reporting systems (i.e., SCADA systems). Note that this discussion does not focus on alarm systems that are not related to a facility's processes.

Alarm systems can be integrated with fire detection systems, intrusion detection systems (IDSs), access control systems, or closed circuit television (CCTV) systems such that these systems automatically respond when the alarm is triggered. For example, a smoke detector alarm can be set up to automatically notify the fire department when smoke is detected; or an intrusion alarm can automatically trigger cameras to turn on in a remote location so that personnel can monitor that location.

An alarm system consists of sensors that detect different types of events; an arming station that is used to turn the system on and off; a control panel that receives information, processes it, and transmits the alarm; and an annunciator that generates a visual and/or audible response to the alarm. When a sensor is tripped, it sends a signal to a control panel, which triggers a visual or audible alarm and/or notifies a central monitoring station. A more complete description of each of the components of an alarm system is provided next.

Detection devices (also called *sensors*) are designed to detect a specific type of event (such as smoke, intrusion, etc.). Depending on the type of event they are designed to detect, sensors can be located inside or outside of the facility or other asset. When an event is detected, the sensors use some type of communication method (such as wireless radio transmitters, conductors, or cables) to send signals to the control panel to generate the alarm. For example, a smoke detector sends a signal to a control panel when it detects smoke.

Alarms use either normally closed (NC) or normally open (NO) electric loops, or "circuits," to generate alarm signals. These two types of circuits are discussed separately below.

In NC loops or circuits, all of the system's sensors and switches are connected in series. The contacts are "at rest" in the closed (on) position, and current continually passes through the system. However, when an event triggers the sensor, the loop is opened, breaking the flow of current through the system and triggering the alarm. NC switches are used more often than are NO switches because the alarm will be activated if the loop or circuit is broken or cut, thereby reducing the potential for circumventing the alarm. This is known as a "supervised" system.

In NO loops or circuits, all of the system's sensors and switches are connected in parallel. The contacts are "at rest" in the open (off) position, and no current passes through the system. However, when an event triggers the sensor, the loop is closed. This allows current to flow through the loop, powering the alarm. NO systems are not "supervised" because the alarm will not be activated if the loop or circuit is broken or cut. However, adding an end-of-line resistor to an NO loop will cause the system to alarm if tampering is detected.

An *arming station*, which is the main user interface with the security system, allows the user to arm (turn on), disarm (turn off), and communicate with the system. How a specific system is armed will depend on how it is used. For example, while IDSs can be armed for continuous operation (twenty-four hours/day), they are usually armed and disarmed according to the work schedule at a specific location so that personnel going about their daily activities do not set off the alarms. In contrast, fire protection systems are typically armed twenty-four hours/day.

A *control panel* receives information from the sensors and sends it to an appropriate location, such as to a central operations station or to a twenty-four-hour monitoring facility. Once the alarm signal is received at the central monitoring location, personnel monitoring for alarms can respond (such as by sending security teams to investigate or by dispatching the fire department).

An *annunciator* responds to the detection of an event by emitting a signal. This signal may be visual, audible, electronic, or a combination of these three. For example, fire alarm signals will always be connected to audible annunciators, whereas intrusion alarms may not be.

Alarms can be reported locally, remotely, or both locally and remotely. Local and remotely (centrally) reported alarms are discussed in more detail below.

A *local alarm* emits a signal at the location of the event (typically using a bell or siren). A "local only" alarm emits a signal at the location of the event but does not transmit the alarm signal to any other location (i.e., it does not transmit the alarm to a central monitoring location). Typically, the purpose of a "local only" alarm is to frighten away intruders and possibly to attract the attention of someone who might notify the proper authorities. Because no signal is sent to a central monitoring location, personnel can respond to a local alarm only if they are in the area and can hear and/or see the alarm signal.

Fire alarm systems must have local alarms, including both audible and visual signals. Most fire alarm signal and response requirements are codified in the National Fire Alarm Code, National Fire Protection Association (NFPA) 72. NFPA 72 discusses the application, installation, performance, and maintenance of protective signaling systems and their components. In contrast to fire alarms, which require a local signal when fire is detected, many IDSs do not have a local alert device because monitoring personnel do not wish to inform potential intruders that they have been detected. Instead, these types of systems silently alert monitoring personnel that an intrusion has been detected, thus allowing monitoring personnel to respond.

In contrast to systems that are set up to transmit "local only" alarms when the sensors are triggered, systems can also be set up to transmit signals to a *central location*, such as a control room or guard post at the utility, or a police or fire station. Most fire/smoke alarms are set up to signal both at the location of the event and at a fire station or central monitoring station. Many insurance companies require that facilities install certified systems that include alarm communication to a central station. For example, systems certified by the Underwriters Laboratory (UL) require that the alarm be reported to a central monitoring station.

The main differences between alarm systems lie in the types of event detection devices used in different systems. Intrusion sensors, for example, consist of two main categories: perimeter sensors and interior (space) sensors. *Perimeter intrusion sensors* are typically applied on fences, doors, walls, windows, and so on, and are designed to detect an intruder before he/she accesses a protected asset (i.e., perimeter intrusion sensors are used to detect intruders attempting to enter through a door, window, etc.). In contrast, *interior intrusion sensors* are designed to detect an intruder who has already accessed the protected asset (i.e., interior intrusion sensors are used to detect intruders once they are already within a protected room or building). These two types of detection devices can be complementary, and they are often used together to enhance security for an asset. For example, a typical intrusion alarm system might employ a perimeter glass-break detector that protects against intruders accessing a room through a window, as well as an ultrasonic interior sensor that detects intruders that have gotten into the room without using the window. Table 9.6 lists and describes types of perimeter and interior sensors.

Fire detection/fire alarm systems consist of different types of fire detection devices and fire alarm systems available. These systems may detect fire, heat, smoke, or a combination of any of these. For example, a typical fire alarm system might consist of heat sensors, which are located throughout a facility and which detect high temperatures or a certain change in temperature over a fixed time period. A different system might be outfitted with both smoke and heat detection devices. A summary of several different types of fire/smoke/heat detection sensors is provided in table 9.7.

Once a sensor in an alarm system detects an event, it must communicate an alarm signal. The two basic types of alarm communication systems are hardwired and wireless. Hardwired systems rely on wire that is run from the control panel to each of the detection devices and annunciators. Wireless systems transmit signals from a transmitter to a receiver through the air—primarily using radio or other waves. Hardwired systems are usually lower in cost, more reliable (they are not affected by terrain or environmental factors), and significantly easier to troubleshoot than are wireless systems. However, a major disadvantage of hardwired systems is that it may not be possible to hardwire all locations (for example, it may be difficult to hardwire remote locations). In addition, running wires to their required locations can be both time consuming and costly. The major advantage to using wireless systems is that they can often be installed in areas where hardwired systems are not feasible. However, wireless components can be much more expensive when compared with hardwired systems. In addition, in the

Table 9.6. Perimeter and Interior Sensors

Type of Perimeter Sensor	Description
Foil	Foil is a thin, fragile, lead-based metallic tape that is applied to glass windows and doors. The tape is applied to the window or door, and electric wiring connects this tape to a control panel. The tape functions as a conductor and completes the electric circuit with the control panel. When an intruder breaks the door or window, the fragile foil breaks, opening the circuit and triggering an alarm condition.
Magnetic switches (reed switches)	The most widely used perimeter sensor. They are typically used to protect doors, as well as windows that can be opened (windows that cannot be opened are more typically protected by foil alarms).
Glass-break detectors	Placed on glass and sense vibrations in the glass when it is disturbed. The two most common types of glass-break detectors are shock sensors and audio discriminators.

Type of Interior Sensor	Description
Passive infrared (PIR)	Presently the most popular and cost-effective interior sensors. PIR detectors monitor infrared radiation (energy in the form of heat) and detect rapid changes in temperature within a protected area. Because infrared radiation is emitted by all living things, these types of sensors can be very effective.
Quad PIRs	Consist of two dual-element sensors combined in one housing. Each sensor has a separate lens and a separate processing circuitry, which allows each lens to be set up to generate a different protection pattern.
Ultrasonic detectors	Emit high-frequency sound waves and sense movement in a protected area by sensing changes in these waves. The sensor emits sound waves that stabilize and set a baseline condition in the area to be protected. Any subsequent movement within the protected area by a would-be intruder will cause a change in these waves, thus creating an alarm condition.
Microwave detectors	Emit ultra-high-frequency radio waves, and the detector senses any changes in these waves as they are reflected throughout the protected space. Microwaves can penetrate through walls, and thus a unit placed in one location may be able to protect multiple rooms.
Dual-technology devices	Incorporate two different types of sensor technology (such as PIR and microwave technology) together in one housing. When both technologies sense an intrusion, an alarm is triggered.

Source: U.S. Environmental Protection Agency, *Water and Wastewater Security Product Guide*, http://cfpub.epa.gov/ safewater/watersecurity/guide (accessed April 4, 2009).

past, it has been difficult to perform self-diagnostics on wireless systems to confirm that they are communicating properly with the controller. Currently, the majority of wireless systems incorporate supervising circuitry, which allows the subscriber to know immediately if there is a problem with the system (such as a broken detection device or a low battery) or if a protected door or window has been left open.

Table 9.7. Fire/Smoke/Heat Detection Sensors

Detector Type	Description
Thermal detectors	Sense when temperatures exceed a set threshold (fixed temperature detectors) or when the rate of change of temperature increases over a fixed time period (rate-of-rise detectors).
Duct detectors	Are located within the heating and ventilation ducts of the facility. This sensor detects the presence of smoke within the system's return or supply ducts. A sampling tube can be added to the detector to help span the width of the duct.
Smoke detectors	Sense invisible and/or visible products of combustion. The two principle types of smoke detectors are photoelectric and ionization detectors. The major differences between these devices are described below: • Photoelectric smoke detectors react to visible particles of smoke. These detectors are more sensitive to the cooler smoke with large smoke particles that is typical of smoldering fires. • Ionization smoke detectors are sensitive to the presence of ions produced by the chemical reactions that take place with few smoke particle, such as those typically produced by fast burning/flaming fires.
Multisensor detectors	Are a combination of photoelectric and thermal detectors. The photoelectric sensor serves to detect smoldering fires, while the thermal detector senses the heat given off from fast burning/flaming fires.
Carbon monoxide (CO) detectors	Are used to indicate the outbreak of fire by sensing the level of carbon monoxide in the air. The detector has an electrochemical cell that senses carbon monoxide but not some other products of combustion.
Beam detectors	Are designed to protect large, open spaces such as industrial warehouses. These detectors consist of three parts: the transmitter, which projects a beam of infrared light; the receiver, which registers the light and produces an electrical signal; and the interface, which processes the signal and generates alarm of fault signals. In the event of a fire, smoke particles obstruct the beam of light. Once a preset threshold is exceeded, the detector will go into alarm.
Flame detectors	Sense either ultraviolet (UV) or infrared (IR) radiation emitted by a fire.
Air-sampling detectors	Actively and continuously sample the air from a protected space and are able to sense the precombustion stages of incipient fire.

Source: U.S. Environmental Protection Agency, *Water and Wastewater Security Product Guide*, http://cfpub.epa.gov/safewater/watersecurity/guide (accessed April 4, 2009).

Backflow Prevention Devices

Backflow prevention devices are designed to prevent backflow, which is the reversal of the normal and intended direction of water flow in a water system. Backflow is a potential problem in a nuclear, chemical, or waste-processing system because if incorrectly cross-connected to potable water it can spread contaminated water back through a distribution system. For example, backflow at uncontrolled cross-connections (cross-connections are any actual or potential connection between the public water supply and a source of chemical contamination) or pollution can allow pollutants or contaminants to enter the potable water system. More specifically, backflow from private plumbing systems, industrial areas, hospitals, and other hazardous contaminant-containing systems into public water mains and wells poses serious public health risks and security problems. Cross-contamination from private plumbing systems can contain biological hazards (such as bacteria or viruses) or toxic substances that can contaminate and sicken an entire population in the event of backflow. The majority of historical incidences of backflow have been accidental, but growing concern that contaminants could be intentionally backfed into a system is prompting increased awareness for private homes, businesses, industries, and areas most vulnerable to intentional strikes. Therefore, backflow prevention is a major tool for the protection of water systems.

Backflow may occur under two types of conditions: backpressure and backsiphonage. *Backpressure* is the reverse from normal flow direction within a piping system that is the result of the downstream pressure being higher than the supply pressure. These reductions in the supply pressure occur whenever the amount of water being used exceeds the amount of water supplied, such as during water line flushing, firefighting, or breaks in water mains. *Backsiphonage* is the reverse from normal flow direction within a piping system that is caused by negative pressure in the supply piping (i.e., the reversal of normal flow in a system caused by a vacuum or partial vacuum within the water supply piping). Backsiphonage can occur where there is a high velocity in a pipe line; when there is a line repair or break that is lower than a service point; or when there is lowered main pressure due to high water withdrawal rate, such as during firefighting or water main flushing.

To prevent backflow, various types of backflow preventers are appropriate for use. The primary types of backflow preventers are

- Air gap drains
- Double check valves
- Reduced pressure principle assemblies
- Pressure vacuum breakers

Biometric Security Systems

Biometrics involves measuring the unique physical characteristics or traits of the human body. Any aspect of the body that is measurably different from person to person—for example, fingerprints or eye characteristics—can serve as a unique biometric identifier for that individual. Biometric systems recognizing fingerprints, palm shape, eyes, face, voice, and signature comprise the bulk of the current biometric systems, but systems that recognize other biological features do exist.

Biometric security systems use biometric technology combined with some type of locking mechanisms to control access to specific assets. In order to access an asset controlled by a biometric security system, an individual's biometric trait must be matched with an existing profile stored in a database. If there is a match between the two, the locking mechanisms (which could be a physical lock, such as at a doorway, an electronic lock, such as at a computer terminal, or some other type of lock) is disengaged, and the individual is given access to the asset.

A biometric security system typically comprises the following components:

- A sensor, which measures/records a biometric characteristic or trait
- A control panel, which serves as the connection point between various system components. The control panel communicates information back and forth between the sensor and the host computer and controls access to the asset by engaging or disengaging the system lock based on internal logic and information from the host computer
- A host computer, which processes and stores the biometric trait in a database
- Specialized software, which compares an individual image taken by the sensor with a stored profile or profiles
- A locking mechanism, which is controlled by the biometric system
- A power source to power the system

Biometric Hand and Finger Geometry Recognition

Hand and finger geometry recognition is the process of identifying an individual through the unique "geometry" (shape, thickness, length, width, etc.) of that individual's hand or fingers. Hand geometry recognition has been employed since the early 1980s and is among the most widely used biometric technologies for controlling access to important assets. It is easy to install and use and is appropriate for use in any location requiring use of two-finger, highly accurate, nonintrusion biometric security. For example, it is currently used in numerous workplaces, day-care facilities, hospitals, universities, airports, refineries, and power plants.

A newer option within hand geometry recognition technology is finger geometry recognition (not to be confused with fingerprint recognition). Finger geometry recognition relies on the same scanning methods and technologies as does hand geometry recognition, but the scanner scans only two of the user's fingers as opposed to his entire hand. Finger geometry recognition has been in commercial use since the mid-1990s and is mainly used in time and attendance applications (i.e., to track when individuals have entered and exited a location). To date the only large-scale commercial use of two-finger geometry for controlling access is at Disney World, where season pass holders use the geometry of their index and middle finger to gain access to the facilities.

To use a hand or finger geometry unit, an individual presents his or her hand or fingers to the biometric unit for "scanning." The scanner consists of a charged coupled device (CCD), which is essentially a high-resolution digital camera, a reflective platen on which the hand is placed, and a mirror or mirrors that help capture different angles of the hand or fingers. The camera "scans" individual geometric characteristics of the hand or fingers by taking multiple images while the user's hand rests on the reflective platen. The camera also captures "depth," or three-dimensional information, through light reflected from the mirrors and the reflective platen. This live image is then compared to a "template" that was previously established for that individual when the person was "enrolled" in the system. If the live scan of the individual matches the stored template, the individual is "verified" and is given access to that asset. Typically, verification takes about two seconds. In access control applications, the scanner is usually connected to some sort of electronic lock, which unlocks the door, turnstile, or other entry barrier when the user is verified. The user can then proceed through the entrance. In time and attendance applications, the time that an individual checks in and out of a location is stored for later use.

As discussed above, hand and finger geometry recognition systems can be used in several types of applications, including access control and time and attendance tracking. While time and attendance tracking can be used for security, it is primarily used for operations and payroll purposes (i.e., clocking in and clocking out). In contrast, access control applications are more likely to be security related. Biometric systems are widely used for access control and can be used on various types of assets, including entryways, computers, vehicles, and so on. However, because of their size, hand/finger recognition systems are primarily used in entryway access control applications.

Biometric Overview—Iris Recognition

The iris, which is the colored or pigmented area of the eye surrounded by the sclera (the white portion of the eye), is a muscular membrane that controls the amount of light entering the eye by contracting or expanding the pupil (the dark center of the eye). The dense, unique patterns of connective tissue in the human iris were first noted in 1936, but it was not until 1994, when algorithms for iris recognition were created and patented, that commercial applications using biometric iris recognition began to be used extensively. There are now two vendors producing iris recognition technology: both the original developer of these algorithms as well as a second company, which has developed and patented a different set of algorithms for iris recognition.

The iris is an ideal characteristic for identifying individuals because it is formed *in utero*, and its unique patterns stabilize around eight months after birth. No two irises are alike; neither an individual's right or left irises, nor the irises of identical twins. The iris is protected by the cornea (the clear covering over the eye), and therefore it is not subject to the aging or physical changes (and potential variation) that are common to some other biometric measures, such as the hand, fingerprints, and the face. Although some limited changes can occur naturally over time, these changes generally occur in the iris's melanin and therefore affect only the eye's color and not its unique patterns (in addition, because iris scanning uses only black-and-white images, color changes would not affect the scan anyway). Thus, barring specific injuries or certain surgeries directly affecting the iris, the iris's unique patterns remain relatively unchanged over an individual's lifetime.

Iris recognition systems employ a monochromatic or black-and-white video camera that uses both visible and near infrared light to take video of an individual's iris. Video is used rather than still photography as an extra security procedure. The video is used to confirm the normal continuous fluctuations of the pupil as the eye focuses, which ensures that the scan is of a living human being, not a photograph or some other attempted hoax. A high-resolution image of the iris is then captured or extracted from the video, using a device often referred to as a "frame grabber." The unique characteristics identified in this image are then converted into a numeric code, which is stored as a "template" for that user.

Card Identification/Access/Tracking Systems

A card-reader system is a type of electronic identification system that is used to identify a card and then perform an action associated with that card. Depending on the system, the card may identify where a person is or was at a certain time, or it may authorize another action, such as disengaging a lock. For example, a security guard may use his card at card readers located throughout a facility to indicate that he has checked a certain location at a certain time. The reader will store the information and/ or send it to a central location, where it can be checked later to ensure that the guard has patrolled the area. Other card-reader systems can be associated with a lock so that the cardholder must have his card read and accepted by the reader before the lock disengages.

A complete card-reader system typically consists of the following components:

- Access cards that are carried by the user
- Card readers, which read the card signals and send the information to control units
- Control units, which control the response of the card reader to the card
- A power source

A "card" may be a typical card or another type of device, such as a key fob or wand. These cards store electronic information, which can range from a simple code (e.g., the alphanumeric code on a Proximity card) to individualized personal data (e.g., biometric data on a smart card). The card reader reads the information stored on the card and sends it to the control unit, which determines the appropriate action to take when a card is presented. For example, in a card access system, the control unit compares the information on the card with stored access authorization information to determine whether the card holder is authorized to proceed through the door. If the information stored in the card-reader system indicates that the key is authorized to allow entrance through the doorway, the system disengages the lock, and the key holder can proceed through the door.

There are many types of card-reader systems on the market. The primary differences between card-reader systems are in the way that data is encoded on the cards, the way these data are transferred between the card and the card reader, and the types of applications for which they are best suited. However, all card systems are similar in the way that the card reader and control unit interact to respond to the card.

While card readers are similar in the way that the card reader and control unit interact to control access, they are different in the way data is encoded on the cards and the way these data are transferred between the card and the card reader. There are several types of technologies available for card reader systems. These include

- Proximity
- Wiegand
- Smartcard
- Magnetic Stripe
- Bar Code
- Infrared
- Barium Ferrite
- Hollerith
- Mixed Technologies

Table 9.8 summarizes various aspects of card-reader technologies. The determination for the level of security rate (low, moderate, or high) is based on the level of technology a given card-reader system has and how simple it is to duplicate that technology and thus bypass the security. Vulnerability ratings were based on whether the card reader can be damaged easily due to frequent use or difficult working conditions (e.g., weather conditions if the reader is located outside). Often this is influenced by the number of moving parts in the system—the more moving parts, the greater the system's potential susceptibility to damage. The life cycle rating is based on the durability of a given card-reader system over its entire operational period. Systems requiring frequent physical contact between the reader and the card often have a shorter life cycle due to the wear and tear to which the equipment is exposed. For many card

Table 9.8. Card-Reader Technology

Types of Card Readers	Technology	Life Cycle	Vulnerability	Level of Security
Proximity	Embedded radio frequency circuits encoded with unique information	Long	Virtually none	Moderate–High
Wiegand	Short lengths of small-diameter, special alloy wire with unique magnetic properties	Long	Low susceptibility to damage; high durability due to embedded wires	Moderate–Expensive
Magnetic Stripe	Electromagnetic charges to encode information on a piece of tape attached to back of card	Moderate	Moderately susceptible to damage due to frequency of use	Low–Moderate
Bar Code	Series of narrow and wide bars and spaces	Short	High; easily damaged.	Low
Hollerith	Holes punched in a plastic or paper card and read optically	Short	High; easily damaged from frequent use	Low
Infrared	An encoded shadow pattern within the card, read using an infrared scanner	Moderate	IR scanners are optical and thus vulnerable to contamination	High
Barium Ferrite	Uses small bits of magnetized barium ferrite placed inside a plastic card; the polarity and location of the "spots" determines the coding	Moderate	Low susceptibility to damage; durable since spots are embedded in the material	Moderate–High
Smart cards	Patterns or series of narrow and wide bars and spaces	Short	High susceptibility to damage, low durability	Highest

Source: U.S. Environmental Protection Agency, *Water and Wastewater Security Product Guide*, http://cfpub.epa.gov/safewater/watersecurity/guide (accessed April 4, 2009).

reader systems, the vulnerability rating and life cycle ratings have a reciprocal relationship. For instance, if a given system has a high vulnerability rating it will almost always have a shorter life cycle.

Card reader technology can be implemented for facilities of any size and with any number of users. However, because individual systems vary in the complexity of their technology and in the level of security they can provide to a facility, individual users must determine the appropriate system for their needs. Some important features to consider when selecting a card-reader system include

- The technological sophistication and security level of the card system
- The size and security needs of the facility
- The frequency with which the card system will be used. For systems that will experience a high frequency of use it is important to consider a system that has a longer life cycle and lower vulnerability rating, thus making it more cost effective to implement.
- The conditions in which the system will be used (i.e., will it be used on the interior or exterior of buildings, does it require light or humidity controls, etc.). Most card-reader systems can operate under normal environmental conditions; therefore, this would be a mitigating factor only in extreme conditions.
- System costs

Exterior Intrusion—Buried Sensors

Buried sensors are electronic devices that are designed to detect potential intruders. The sensors are buried along the perimeters of sensitive assets and are able to detect intruder activity both above- and belowground. Some of these systems are composed of individual stand-alone sensor units, while other sensors consist of buried cables.

There are four types of buried sensors that rely on different types of triggers. These are pressure or seismic, magnetic field, ported coaxial cable, and fiber-optic cables.

Table 9.9. Types of Buried Sensors

Type	Description
Pressure or seismic	Responds to disturbances in the soil
Magnetic field	Responds to a change in the local magnetic field caused by the movement of nearby metallic material
Ported coaxial cables	Responds to motion of a material with a high dielectric constant or high conductivity near the cables
Fiber-optic cables	Responds to a change in the shape of the fiber that can be sensed using sophisticated sensors and computer signal processing

Source: Adapted from M. L. Garcia (2001), *The Design and Evaluation of Physical Protection Systems*. Boston: Butterworth-Heinemann.

These four sensors are all covert and terrain following, meaning they are hidden from view and follow the contour of the terrain. The four types of sensors are described in more detail below. Table 9.9 presents the distinctions between the four types of buried sensors.

Exterior Intrusion Sensors

An exterior intrusion sensor is a detection device that is used in an outdoor environment to detect intrusions into a protected area. These devices are designed to detect an intruder and then communicate an alarm signal to an alarm system. The alarm system can respond to the intrusion in different ways, such as by triggering an audible or visual alarm signal or by sending an electronic signal to a central monitoring location that notifies security personnel of the intrusion.

Intrusion sensors can be used to protect many kinds of assets. Intrusion sensors that protect physical space are classified according to whether they protect indoor, or "interior" space (i.e., an entire building or room within a building), or outdoor, or "exterior" space (i.e., a fence line or perimeter). Interior intrusion sensors are designed to protect the interior space of a facility by detecting an intruder who is attempting to enter or who has already entered a room or building. In contrast, exterior intrusion sensors are designed to detect an intrusion into a protected outdoor/exterior area. Exterior protected areas are typically arranged as zones or exclusion areas placed so that the intruder is detected early in the intrusion attempt before the intruder can gain access to more valuable assets (e.g., into a building located within the protected area). Early detection creates additional time for security forces to respond to the alarm.

Exterior intrusion sensors are classified according to how the sensor detects the intrusion within the protected area. The three classes of exterior sensor technology include

- Buried-line sensors
- Fence-associated sensors
- Freestanding sensors

1. Buried-line sensors—As the name suggests, buried-line sensors are sensors that are buried underground and are designed to detect disturbances within the ground— such as disturbances caused by an intruder digging, crawling, walking, or running on the monitored ground. Because they sense ground disturbances, these types of sensors are able to detect intruder activity both on the surface and belowground. Individual types of exterior buried-line sensors function in different ways, including by detecting motion, pressure, or vibrations within the protected ground or by detecting changes in some type of field (e.g., magnetic field) that the sensors

generate within the protected ground. Specific types of buried-line sensors include pressure or seismic sensors, magnetic field sensors, ported coaxial cables, and fiber-optic cables. Details on each of these sensor types are provided below.

- *Buried-line pressure* or *seismic sensors* detect physical disturbances to the ground—such as vibrations or soil compression—caused by intruders walking, driving, digging, or otherwise physically contacting the protected ground. These sensors detect disturbances from all directions and, therefore, can protect an area radially outward from their location; however, because detection may weaken as a function of distance from the disturbance, choosing the correct burial depth from the design area will be crucial. In general, sensors buried at a shallow depth protect a relatively small area but have a high probability of detecting intrusion within that area, while sensors buried at a deeper depth protect a wider area but have a lower probability of detecting intrusion into that area.

- *Buried-line magnetic field sensors* detect changes in a local magnetic field that are caused by the movement of metallic objects within that field. This type of sensor can detect ferric metal objects worn or carried by an intruder entering a protected area on foot as well as vehicles being driven into the protected area.

- *Buried-line ported coaxial cable sensors* detect the motion of any object (i.e., human body, metal, etc.) possessing high conductivity and located within close proximity to the cables. An intruder entering into the protected space creates an active disturbance in the electric field, thereby triggering an alarm condition.

- *Buried-line fiber-optic cable sensors* detect changes in the attenuation of light signals transmitted within the cable. When the soil around the cable is compressed, the cable is distorted, and the light signal transmitted through the cable changes, initiating an alarm. This type of sensor is easy to install because it can be buried at a shallow depth (only a few centimeters) and still be effective.

2. Fence-associated sensors—Fence-associated sensors are either attached to an existing fence or are installed in such a way as to create a fence. These sensors detect disturbances to the fence—such as those caused by an intruder attempting to climb the fence or by an intruder attempting to cut or lift the fence fabric. Exterior fence-associated sensors include fence-disturbance sensors, taut-wire sensor fences, and electric field or capacitance sensors. Details on each of these sensor types are provided below.

   - *Fence-disturbance sensors* detect the motion or vibration of a fence, such as that caused by an intruder attempting to climb or cut through the fence. In general, fence disturbance sensors are used on chain link fences or on other fence types where a movable fence fabric is hung between fence posts.

   - *Taut-wire sensor fences* are similar to fence-disturbance sensors except that instead of attaching the sensors to a loose fence fabric, the sensors are attached to a wire that is stretched tightly across the fence. These types of systems are designed to detect changes in the tension of the wire rather than vibrations in the fence fabric. Taut-wire sensor fences can be installed over existing fences or as stand-alone fence systems.

- *Electric field or capacitance sensors* detect changes in capacitive coupling between wires that are attached to, but electrically isolated from, the fence. As opposed to other fence-associated intrusion sensors, both electric field and capacitance sensors generate an electric field that radiates out from the fence line, resulting in an expanded zone of protection relative to other fence-associated sensors and allowing the sensor to detect intruders' presence before they arrive at the fence line. Note: Proper spacing is necessary during installation of the electric field sensor to detect a would-be intruder from slipping between largely spaced wires.

3. *Freestanding Sensors*—These sensors, which include active infrared, passive infrared, bistatic microwave, monostatic microwave, dual-technology, and video motion detection (VMD) sensors, consist of individual sensor units or components that can be set up in a variety of configurations to meet a user's needs. They are installed aboveground, and depending on how they are oriented relative to each other, they can be used to establish a protected perimeter or a protected space. More details on each of these sensor types are provided below.

- *Active infrared sensors* transmit infrared energy into the protected space and monitor for changes in this energy caused by intruders entering that space. In a typical application, an infrared light beam is transmitted from a transmitter unit to a receiver unit. If an intruder crosses the beam, the beam is blocked, and the receiver unit detects a change in the amount of light received, triggering an alarm. Different sensors can see single- and multiple-beam arrays. Single-beam infrared sensors transmit a single infrared beam. In contrast, multiple-beam infrared sensors transmit two or more beams parallel to each other. This multiple-beam sensor arrangement creates an infrared "fence."

- *Passive infrared (PIR) sensors* monitor the ambient infrared energy in a protected area and evaluate changes in that ambient energy that may be caused by intruders moving through the protected area. Detection ranges can exceed one hundred yards on cold days with size and distance limitations dependent upon the background temperature. PIR sensors generate a nonuniform detection pattern (or "curtain") that has areas (or "zones") of more sensitivity and areas of less sensitivity. The specific shape of the protected area is determined by the detector's lenses. The general shape common to many detection patterns is a series of long "fingers" emanating from the PIR and spreading in various directions. When intruders enter the detection area, the PIR sensor detects differences in temperature due to the intruder's body heat and triggers an alarm. While the PIR leaves unprotected areas between its fingers, an intruder would be detected if he passed from a nonprotected area to a protected area.

- *Microwave sensors* detect changes in received energy generated by the motion of an intruder entering into a protected area. Monostatic microwave sensors incorporate a transmitter and a receiver in one unit, while bistatic sensors separate the transmitter and the receiver into different units. Monostatic sensors are limited to a coverage area of four hundred feet, while bistatic sensors can cover an area of up to fifteen hundred feet. For bistatic sensors, a zone of no detection exists in the first few feet in front of the antennas. This distance from the antennas to the point at which the intruder is first detected is known as the offset distance. Due to this offset distance, antennas must be configured so that they overlap one another (as opposed to being adjacent to one another), thereby creating long perimeters with a continuous line of detection.

- *Dual-technology sensors* consist of two different sensor technologies incorporated together into one sensor unit. For example, a dual technology sensor could consist of a passive infrared detector and a monostatic microwave sensor integrated into the same sensor unit.

- *Video motion detection (VMD) sensors* monitor video images from a protected area for changes in the images. Video cameras are used to detect unauthorized intrusion into the protected area by comparing the most recent image to a previously established one. Cameras can be installed on towers or other tall structures so that they can monitor a large area.

Fences

A fence is a physical barrier that can be set up around the perimeter of an asset. Fences often consist of individual pieces (such as individual pickets in a wooden fence or individual sections of a wrought iron fence) that are fastened together. Individual sections of the fence are fastened together using posts, which are sunk into the ground to provide stability and strength for the sections of the fence hung between them. Gates are installed between individual sections of the fence to allow access inside the fenced area.

Many fences are used as decorative architectural features to separate physical spaces for each other. They may also be used to physically mark the location of a boundary (such as a fence installed along a property line). However, a fence can also serve as an effective means for physically delaying intruders from gaining access to an energy sector asset. For example, many utilities install fences around their primary facilities, around remote pump stations, or around hazardous petrochemical materials storage areas or sensitive areas within a facility. Access to the area can be controlled through security at gates or doors through the fence (for example, by posting a guard at the gate or by locking it). In order to gain access to the asset, unauthorized persons could either have to go around or through the fence.

Fences are often compared with walls when determining the appropriate system for perimeter security. While both fences and walls can provide adequate perimeter security, fences are often easier and less expensive to install than walls. However, they do not usually provide the same physical strength that walls do. In addition, many types of fences have gaps between the individual pieces that make up the fence (i.e., the spaces between chain links in a chain link fence or the space between pickets in a picket fence). Thus, many types of fences allow the interior of the fenced area to be seen. This may allow intruders to gather important information about the locations or defenses of vulnerable areas within the facility.

There are numerous types of materials used to construct fences, including chain link iron, aluminum, wood, or wire. Some types of fences, such as split rails or pickets, may not be appropriate for security purposes because they are traditionally low fences, and they are not physically strong. Potential intruders may be able to easily defeat these fences either by jumping or climbing over them or by breaking through them. For example, the rails in a split fence may be able to break easily.

Important security attributes of a fence include the height to which it can be constructed, the strength of the material comprising the fence, the method and strength of attaching the individual sections of the fence together at the posts and the fence's ability to restrict the view of the assets inside the fence. Additional considerations should include the ease of installing the fence and the ease of removing and reusing sections of the fence. Table 9.10 provides a comparison of the important security and usability features of various fence types.

Some fences can include additional measures to delay, or even detect, potential intruders. Such measures may include the addition of barbed wire, razor wire, or other deterrents at the top of the fence. Barbed wire is sometimes employed at the base of fences as well. This can impede a would-be intruder's progress in even reaching the fence. Fences may also be fitted with security cameras to provide visual surveillance of the perimeter. Finally, some facilities have installed motion sensors along their fences

Table 9.10. Comparison of Different Fence Types

Specifications	Chain Link	Iron	Wire (Wirewall)	Wood
Height limitations	12 feet	12 feet	12 feet	8 feet
Strength	Medium	High	High	Low
Installation requirements	Low	High	High	Low
Ability to remove/Reuse	Low	High	Low	High
Ability to replace/repair	Medium	High	Low	High

Source: U.S. Environmental Protection Agency, *Water and Wastewater Security Product Guide*, http://cfpub.epa.gov/safewater/watersecurity/guide (accessed April 4, 2009).

to detect movement on the fence. Several manufacturers have combined these multiple perimeter security features into one product and offer alarms and other security features.

The correct implementation of a fence can make it a much more effective security measure. Security experts recommend the following when a facility constructs a fence:

- The fence should be at least seven to nine feet high.
- Any outriggers, such as barbed wire, that are affixed on top of the fence should be angled out and away from the facility, and not in toward the facility. This will make climbing the fence more difficult and will prevent ladders from being placed against the fence.
- Other types of hardware can increase the security of the fence. This can include installing concertina wire along the fence (this can be done in front of the fence or at the top of the fence), or adding intrusion sensors, camera, or other hardware to the fence.
- All undergrowth should be cleared for several feet (typically six feet) on both sides of the fence. This will allow for a clearer view of the fence by any patrols in the area.
- Any trees with limbs or branches hanging over the fence should be trimmed so that intruders cannot use them to go over the fence. Also, it should be noted that fallen trees can damage fences, and so management of trees around the fence can be important. This can be especially important in areas where fence goes through a remote area.
- Fences that do not block the view from outside the fence to inside the fence allow patrols to see inside the fence without having to enter the facility.
- "No Trespassing" signs posted along the fence can be a valuable tool in prosecuting any intruders who claim that the fence was broken and that they did not enter through the fence illegally. Adding signs that highlight the local ordinances against trespassing can further persuade simple troublemakers from illegally jumping/climbing the fence. Electrical substation and other electrical component installations should have clearly visible signage warning of HIGH VOLTAGE and the dangers of electrical shock.

Films for Glass Shatter Protection

Most energy sector utilities have numerous windows on the outside of buildings, in doors, and in interior offices. In addition, many facilities have glass doors or other glass structures, such as glass walls or display cases. These glass objects are potentially vulnerable to shattering when heavy objects are thrown or launched at them, when explosions occur near them, or when there are high winds (for exterior glass). If the glass is shattered, intruders may potentially enter an area. In addition, shattered glass projected into a room from an explosion or from an object being thrown through a door or window can injure and potentially incapacitate personnel in the room. Materials that prevent glass from shattering can help to maintain the integrity of the door, window, or other glass object and can delay an intruder from gaining access. These materials can also prevent flying glass and thus reduce potential injuries.

Materials designed to prevent glass from shattering include specialized films and coatings. These materials can be applied to existing glass objects to improve their strength and their ability to resist shattering. The films have been tested against many scenarios that could result in glass breakage, including penetration by blunt objects, bullets, high winds, and simulated explosions. Thus, the films are tested against both simulated weather scenarios (which could include both the high winds themselves and the force of objects blown into the glass), as well as more criminal/terrorist scenarios where the glass is subject to explosives or bullets. Many vendors provide information on the results of these types of tests, and thus potential users can compare different product lines to determine which products best suit their needs.

The primary attributes of films for shatter protection are:

- The materials from which the film is made
- The adhesive that bonds the film to the glass surface
- The thickness of the film

Standard glass safety films are designed from high-strength polyester. Polyester provides both strength and elasticity, which is important in absorbing the impact of an object, spreading the force of the impact over the entire film, and resisting tearing. The polyester is also designed to be resistant to scratching, which can result when films are cleaned with abrasives or other industrial cleaners.

The bonding adhesive is important in ensuring that the film does not tear away from the glass surface. This can be especially important when the glass is broken so that the film does not peel off the glass and allow it to shatter. In addition, films applied to exterior windows can be subject to high concentrations of UV light, which can break down bonding materials.

Film thickness is measured in gauge or mils. According to test results reported by several manufacturers, film thickness appears to affect resistance to penetration/tearing, with thicker films being more resistant to penetration and tearing. However, the appreciation of a thicker film did not decrease glass fragmentation.

Many manufacturers offer films in different thicknesses. The "standard" film is usually one four-mil layer; thicker films are typically composed of several layers of the standard four-mil sheet. However, newer technologies have allowed the polyester to be "microlayered" to produce a stronger film without significantly increasing its thickness. In this microlayering process, each laminate film is composed of multiple microthin layers of polyester woven together at alternating angles. This provides increased strength for the film, while maintaining the flexibility and thin profile of one film layer.

As described above, many vendors test their products in various scenarios that would lead to glass shattering, including simulated bomb blasts and simulation of the glass being struck by windblown debris. Some manufacturers refer to the Government Services Administration standard for bomb blasts, which require resistance to tearing for a four PSI blast. Other manufacturers use other measures and test for resistance to tearing. Many of these tests are not "standard," in that no standard testing or reporting methods have been adopted by any of the accepted standards-setting institutions. However, many of the vendors publish the procedure and the results of these tests on their websites, and this may allow users to evaluate the protectiveness of these films. For example, several vendors evaluate the "protectiveness" of their films and the "hazard" resulting from blasts near windows with and without protective films. Protectiveness is usually evaluated based on the percentage of glass ejected from the window and the height at which that ejected glass travels during the blast (for example, if the blasted glass tends to project upward into a room—potentially toward people's faces—it is a higher hazard than if it is blown downward into the room toward people's feet). There are some standard measures of glass breakage. For example, several vendors indicated that their products exceed the American Society for Testing and Materials (ASTM) standard 64Z-95 "Standard Test Method for Glazing and Glazing Systems Subject to Air Blast Loadings." Vendors often compare the results of some sort of penetration or force test, ballistic tests, or simulated explosions with unprotected glass versus glass onto which their films have been applied. Results generally show that applying films to the glass surfaces reduces breakage/penetration of the glass and can reduce the amount and direction of glass ejected from the frame. This in turn reduces the hazard from flying glass.

In addition to these types of tests, many vendors conduct standard physical tests on their products, such as tests for tensile strength and peel strength. Tensile strength indicates the strength per area of material, while the peel strength indicates the force it would take to peel the product from the glass surface. Several vendors indicate that their products exceed American National Standards Institute (ANSI) standard Z97.1 for tensile strength and adhesion.

Vendors typically have a warranty against peeling or other forms of deterioration of their products. However, the warranty requires that the films be installed by manufacturer-certified technicians to ensure that they are applied correctly and therefore that the warranty is in effect. Warranties from different manufacturers may vary. Some may cover the cost of replacing the material only, while others include material plus installation. Because installation costs are significantly greater than material costs, different warranties may represent large differences in potential costs.

Fire Hydrant Locks

Fire hydrants are installed at strategic locations throughout a community's water distribution system to supply water for firefighting. However, because there are many hydrants in a system and they are often located in residential neighborhoods, industrial districts, and other areas where they cannot be easily observed and/or guarded, they are potentially vulnerable to unauthorized access. Many municipalities, states, and EPA regions have recognized this potential vulnerability and have instituted programs to lock hydrants. For example, EPA Region 1 has included locking hydrants as number seven on its "Drinking Water Security and Emergency Preparedness" Top Ten List for small groundwater suppliers.

A "hydrant lock" is a physical security device designed to prevent unauthorized access to the water supply through a hydrant. They can also ensure water and water pressure availability to firefighters and prevent water theft and associated lost water revenue. These locks have been successfully used in numerous municipalities and in various climates and weather conditions.

Fire hydrant locks are basically steel covers or caps that are locked in place over the operating nut of a fire hydrant. The lock prevents unauthorized persons from accessing the operating nut and opening the fire hydrant valve. The lock also makes it more difficult to remove the bolts from the hydrant and access the system that way. Finally, hydrant locks shield the valve from being broken off. Should a vandal attempt to breach the hydrant lock by force and succeed in breaking the hydrant lock, the vandal will only succeed in bending the operating valve. If the hydrant's operating valve is bent, the hydrant will not be operational, but the water asset remains protected and inaccessible to vandals. However, the entire hydrant will need to be replaced.

Hydrant locks are designed so that the hydrants can be operated by special "key wrenches" without removing the lock. These specialized wrenches are generally distributed to the fire department, public works department, and other authorized persons so that they can access the hydrants as needed. An inventory of wrenches and their serial numbers is generally kept by a municipality so that the location of all wrenches is known. These operating key wrenches may be purchased only by registered lock owners.

The most important features of hydrants are their strength and the security of their locking systems. The locks must be strong so that they cannot be broken off. Hydrant locks are constructed from stainless or alloyed steel. Stainless steel locks are stronger and are ideal for all climates; however, they are more expensive than alloy locks. The locking mechanisms for each fire hydrant locking system ensure that the hydrant can be operated only by authorized personnel who have the specialized key to work the hydrant.

Hatch Security

A hatch is basically a door that is installed on a horizontal plane (such as in a floor, a paved lot, or a ceiling) instead of on a vertical plane (such as in a building wall). Hatches usually provide access to assets that are either located underground (such as hatches to basements or underground vaults and storage areas) or located above ceilings (such as emergency roof exits). At chemical industrial facilities, hatches are typically used to provide access to underground vaults containing pumps, meter chambers, valves, or piping, or to the interior of chemical tanks or covered water reservoirs. Securing a hatch by locking it or upgrading materials to give the hatch added strength can help to delay unauthorized access to any asset behind the hatch.

Like all doors, a hatch consists of a frame anchored to the horizontal structure, a door or doors, hinges connecting the door/doors to the frame, and a latching or locking mechanism that keeps the hatch door/doors closed.

It should be noted that improving hatch security is straightforward and that hatches with upgraded security features can be installed new or they can be retrofit for existing applications.

Depending on the application, the primary security-related attributes of a hatch are the strength of the door and frame, its resistance to the elements and corrosion, its ability to be sealed against water or gas, and its locking features.

Hatches must be both strong and lightweight so that they can withstand typical static loads (such as people or vehicles walking or driving over them) while still being easy to open.

In addition, because hatches are typically installed at outdoor locations, they are usually designed to be made from corrosion-resistant metal that can withstand the elements. Therefore, hatches are typically constructed from high-gauge steel or lightweight aluminum.

Aluminum is typically the material of choice for hatches because it is lightweight and more corrosion resistant relative to steel. Aluminum is not as rigid as steel, so aluminum hatch doors may be reinforced with aluminum stiffeners to provide extra strength and rigidity. The doors are usually constructed from single or double layers (or "leaves") of material. Single-leaf designs are standard for smaller hatches, while double-leaf deigns are required for larger hatches. In addition, aluminum products do not require painting. This is reflected in the warranties available with different products. Product warranties range from ten years to lifetime.

Steel is heavier per square foot than aluminum, and thus steel hatches will be heavier and more difficult to open than aluminum hatches of the same size. However, heavy steel hatch doors may have spring-loaded, hydraulic, or gas openers or other specialized features that help in opening the hatch and in keeping it open.

Many hatches are installed in outdoor areas, often in roadways or pedestrian areas. Therefore, the hatch installed for any given application must be designed to withstand the expected load at that location. Hatches are typically solid to withstand either pedestrian or vehicle loading. Pedestrian loading hatches are typically designed to withstand either 150 or 300 pounds per square feet (psf) of loading. The vehicle loading standard is the American Association of State Highway and Transportation Officials (AASHTO) H-20 wheel loading standard of sixteen thousand pounds over an eight-inch by twenty-inch area. It should be noted that these design parameters are for static loads and not dynamic loads; thus, the loading capabilities may not reflect potential resistance to other types of loads that may be more typical of an intentional threat, such as repeated blows from a sledgehammer or pressure generated by bomb lasts or bullets.

The typical design for a watertight hatch includes a channel frame that directs water away from the hatch. This can be especially important in a hatch on a storage tank because this will prevent liquid contaminants from being dumped on the hatch and leaking through into the interior. Hatches can also be constructed with gasket seals that are air-, odor-, and gas-tight.

Typically, hatches for pedestrian loading applications have hinges located on the exterior of the hatch, while hatches designed for H-20 loads have hinges located in the interior of the hatch. Hinges located on the exterior of the hatch may be able to be removed; thereby allowing intruders to remove the hatch door and access the asset behind the hatch. Therefore, installing H-20 hatches even for applications that do not require H-20 loading levels may increase security because intruders will not be able to tamper with the hinges and circumvent the hatch this way.

In addition to the location of the hinges, stock hinges can be replaced with heavy duty or security hinges that are more resistant to tampering.

The hatch locking mechanism is perhaps the most important part of hatch security. A number of locks can be implemented for hatches, including

- Slam locks (internal locks that are located within the hatch frame)
- Recessed cylinder locks
- Bolt locks
- Padlocks

Ladder Access Control

Nuclear power plant sector facilities have a number of assets that are raised above ground level, including electrical substations, pumping stations, gas compression facilities, raised petrochemical tanks, raised piping systems, and roof access points into buildings. In addition, communications equipment, antennas, or other electronic devices may be located on the top of these raised assets. Typically, these assets are reached by ladders that are permanently anchored to the asset. For example, raised petrochemical/water tanks typically are accessed by ladders that are bolted to one of the legs of the tank. Controlling access to these raised assets by controlling access to the ladder can increase security at an energy sector facility.

A typical ladder access control system consists of some type of cover that is locked or secured over the ladder. The cover can be a casing that surrounds most of the ladder, or a door or shield that covers only part of the ladder. In either case, several rungs of the ladder (the number of rungs depends on the size of the cover) are made inaccessible by the cover, and these rungs can be accessed only by opening or removing the cover. The cover is locked so that only authorized personnel can open or remove it and use the ladder. Ladder access controls are usually installed at several feet above ground level, and they usually extend several feet up the ladder so that they cannot be circumvented by someone accessing the ladder above the control system.

The important features of ladder access control are the size and strength of the cover and its ability to lock or otherwise be secured from unauthorized access.

The covers are constructed from aluminum or some type of steel. This should provide adequate protection from being pierced or cut through. The metals are corrosion resistant so that they will not corrode or become fragile from extreme weather conditions in outdoor applications. The bolts used to install each of these systems are galvanized steel. In addition, the bolts for each cover are installed on the inside of the unit so that they cannot be removed from the outside.

Locks

A lock is a type of physical security device that can be used to delay or prevent a door, a gate, a window, a manhole, a filing cabinet drawer, or some other physical feature from being opened, moved, or operated. Locks typically operate by connecting two pieces together—such as by connecting a door to a door jamb or a manhole to its casement. Every lock has two modes—engaged (or "locked") and disengaged (or "opened"). When a lock is disengaged, the asset on which the lock is installed can be accessed by anyone, but when the lock is engaged, only those with keys can gain access to the locked asset.

Before discussing locks and their applicability, it is important to discuss key control. Based on our experience, many critical infrastructure facilities (and others) have no idea how many keys for various site/equipment locks have been issued to employees over the years. Many facilities simply issue keys to employees at hiring with no accountability for the keys upon the employee's departure. Needless to say, this is not good security policy. You can have the best-made locks available installed throughout your facilities, but if you do not have proper key control, you do not have proper security.

Locks are excellent security features because they have been designed to function in many ways and to work on many different types of assets. Locks can also provide different levels of security, depending on how they are designed and implemented. The security provided by a lock is dependent on several factors, including its ability to withstand physical damage (i.e., can it be cut off, broken, or otherwise physically disabled) as well as its requirements for supervision or operation (i.e., combinations may need to be changed frequently so that they are not compromised and the locks remain secure). While there is no single definition of the "security" of a lock, locks are often described as minimum, medium, or maximum security. Minimum-security locks are those that can be easily disengaged (or "picked") without the correct key or code or those that can be disabled easily (such as small padlocks that can be cut with bolt cutters). Higher-security locks are more complex and thus are more difficult to pick, or they are sturdier and more resistant to physical damage.

Many locks, such as many door locks, need to be unlocked from only one side. For example, most door locks need a key to be unlocked only from the outside. A person opens such devices, called single-cylinder locks, from the inside by pushing a button or by turning a knob or handle. Double-cylinder locks require a key to be locked or unlocked from both sides.

Manhole Intrusion Sensors

Manholes are commonly found in energy sector industrial sites. Manholes are designed to provide access to the underground utilities, meter vaults, petrochemical pumping rooms, and so on, and therefore they are potential entry points to a system. Because many utilities run under other infrastructure (roads, buildings), manholes also provide potential access points to critical infrastructure as well as petrochemical process assets. In addition, because the portion of the system to which manholes provide entry is primarily located underground, access to a system through a manhole increases the chance that an intruder will not be seen. Therefore, protecting manholes can be a critical component of guarding an entire plant site and a surrounding community.

There are multiple methods for protecting manholes, including preventing unauthorized personnel from physically accessing the manhole and detecting attempts at unauthorized access to the manhole.

A manhole intrusion sensor is a physical security device designed to detect unauthorized access to the facility through a manhole. Monitoring a manhole that provides access to a chemical plant or processing system can mitigate two distinct types of threats. First, monitoring a manhole may detect access of unauthorized personnel to chemical systems or assets through the manhole. Second, monitoring manholes may also allow the detection of intruders attempting to place explosive or other destructive (WMD) devices into the petrochemical system.

Several different technologies have been used to develop manhole intrusion sensors, including mechanical systems, magnetic systems, and fiber-optic and infrared sensors. Some of these intrusion sensors have been specifically designed for manholes, while others consist of standard, off-the-shelf intrusion sensors that have been implemented in a system specifically designed for application in a manhole.

Manhole Locks

A "manhole lock" is a physical security device designed to delay unauthorized access to the energy sector facility or system through a manhole.

Radiation Detection Equipment for Monitoring Personnel and Packages

One of the major potential threats facing any critical production facility or system is contamination by radioactive substances. Radioactive substances brought on-site at a facility could be used to contaminate the facility, thereby preventing workers from safely entering the facility to perform necessary tasks. In addition, radioactive substances brought on-site at an energy sector entity could be discharged into the petrochemical waste system, contaminating the downstream water supply. Therefore, detection of radioactive substances being brought on-site can be an important security enhancement.

Different radionuclides have unique properties, and different equipment is required to detect different types of radiation. However, it is impractical and potentially unnecessary to monitor for specific radionuclides being brought on-site. Instead, for security purposes, it may be more useful to monitor for gross radiation as an indicator of unsafe substances.

In order to protect against these radioactive materials being brought on-site, a facility may set up monitoring sites outfitted with radiation detection instrumentation at entrances to the facility. Depending on the specific types of equipment chosen, this equipment would detect radiation emitted from people, packages, or other objects being brought through an entrance.

One of the primary differences between the different types of detection equipment is the means by which the equipment reads the radiation. Radiation may be detected either by direct measurement or through sampling.

Direct radiation measurement involves measuring radiation through an external probe on the detection instrumentation. Some direct measurement equipment detects radiation emitted into the air around the monitored object. Because this equipment detects radiation in the air, it does not require that the monitoring equipment make physical contact with the monitored object. Direct means for detecting radiation include using a walk-through portal-type monitor that would detect elevated radiation levels on a person or in a package, or by using a handheld detector, which would be moved or swept over individual objects to locate a radioactive source.

Reservoir Covers

Reservoirs are used to store raw or untreated water for various purposes. They can be located underground (buried), at ground level, or on an elevated surface. Reservoirs can vary significantly in size; small reservoirs can hold as little as a thousand gallons, while larger reservoirs may hold many millions of gallons.

Reservoirs can be either natural or man-made. Natural reservoirs can include lakes or other contained water bodies, while man-made reservoirs usually consist of some sort of engineered structure, such as a tank or other impoundment structure. In addition to the water containment structure itself, reservoir systems may also include associated water treatment and distribution equipment, including intakes, pumps, pump houses, piping systems, chemical treatment and chemical storage areas, and so on.

One of the most serious potential threats to the system is direct contamination of the stored water through dumping contaminants into the reservoir. Critical infrastructure facilities have taken various measures to mitigate this type of threat, including fencing off the reservoir, installing cameras to monitor for intruders, and monitoring for changes in water quality. Another option for enhancing security is covering the reservoir using some type of manufactured cover to prevent intruders from gaining physical access to the stored water. Implementing a reservoir cover may or may not be practical, depending on the size of the reservoir (for example, covers are not typically used on natural reservoirs because they are too large for the cover to be technically feasible and cost effective).

A reservoir cover is a structure installed on or over the surface of the reservoir to minimize water quality degradation. The three basic design types for reservoir covers are

- Floating
- Fixed
- Air-supported

A variety of materials are used when manufacturing a cover, including reinforced concrete, steel, aluminum, polypropylene, chlorosulfonated polyethylene, or ethylene interpolymer alloys. There are several factors that affect a reservoir cover's effectiveness, and thus its ability to protect the stored water. These factors include

- The location, size, and shape of the reservoir
- The ability to lay/support a foundation (for example, footing, soil, and geotechnical support conditions)
- The length of time reservoir can be removed from service for cover installation or maintenance
- Aesthetic considerations
- Economic factors, such as capital and maintenance costs

For example, it may not be practical to install a fixed cover over a reservoir if the reservoir is too large or if the local soil conditions cannot support a foundation. A floating or air-supported cover may be more appropriate for these types of applications.

In addition to the practical considerations for installation of these types of covers, there are a number of operations and maintenance (O&M) concerns that affect the utility of a cover for specific applications, including how different cover materials will withstand local climatic conditions, what types of cleaning and maintenance will be required for each particular type of cover, and how these factors will affect the cover's life span and its ability to be repaired when it is damaged.

The primary feature affecting the security of a reservoir cover is its ability to maintain its integrity. Any type of cover, no matter what its construction material, will provide good protection from contamination by rainwater or atmospheric deposition, as well as from intruders attempting to access the stored water with the intent of causing intentional contamination. The covers are large and heavy, and it is difficult to circumvent them to get into the reservoir. At the very least, it would take a determined intruder, as opposed to a vandal, to defeat the cover.

Passive Security Barriers

One of the most basic threats facing any facility is from intruders accessing the facility with the intention of causing damage to its assets. These threats may include intruders actually entering the facility, as well as intruders attacking the facility from outside without actually entering it (i.e., detonating bomb near enough to the facility to cause damage within its boundaries).

Security barriers are one of the most effective ways to counter the threat of intruders accessing a facility or the facility perimeter. Security barriers are large, heavy structures that are used to control access through a perimeter by either vehicles or personnel. They can be used in many different ways, depending on how/where they are located at the facility. For example, security barriers can be used on or along driveways or roads to direct traffic to a checkpoint (i.e., a facility may install jersey barriers in a road to direct traffic in a certain direction). Other types of security barriers (crash beams, gates) can be installed at the checkpoint so that guards can regulate which vehicles can access the facility. Finally, other security barriers (i.e., bollards or security planters) can be used along the facility perimeter to establish a protective buffer area between the facility and approaching vehicles. Establishing such a protective buffer can help in mitigating the effects of the type of bomb blast described above, both by potentially absorbing some of the blast and also by increasing the "stand-off" distance between the blast and the facility (the force of an explosion is reduced as the shock wave travels farther from the source, and thus the farther the explosion is from the target, the less effective it will be in damaging the target).

Security barriers can be either "active" or "passive." "Active" barriers, which include gates, retractable bollards, wedge barriers, and crash barriers, are readily movable, and thus they are typically used in areas where they must be moved often to allow vehicles to pass—such as in roadways at entrances and exits to a facility. In contrast to active security barriers, "passive" security barriers, which include jersey barriers, bollards, and security planters, are not designed to be moved on a regular basis, and thus they are typically used in areas where access is not required or allowed—such as along building perimeters or in traffic control areas. Passive security barriers are typically large, heavy structures that are usually several feet high, and they are designed so that even heavy-duty vehicles cannot go over or through them. Therefore, they can be placed in a roadway parallel to the flow of traffic so that they direct traffic in a certain direction (such as to a guardhouse, a gate, or some other sort of checkpoint), or perpendicular to traffic such that they prevent a vehicle from using a road or approaching a building or area.

Security for Doorways—Side-Hinged Doors

Doorways are the main access points to a facility or to rooms within a building. They are used on the exterior or in the interior of buildings to provide privacy and security for the areas behind them. Different types of doorway security systems may be installed in different doorways depending on the needs or requirements of the buildings or rooms. For example, exterior doorways tend to have heavier doors to withstand the elements and to provide some security to the entrance of the building. Interior doorways in office areas may have lighter doors that may be primarily designed to provide privacy rather than security. Therefore, these doors may be made of glass or lightweight wood. Doorways in industrial areas may have sturdier doors than do other interior doorways and may be designed to provide protection or security for areas behind the doorway. For example, fireproof doors may be installed in chemical storage areas or in other areas where there is a danger of fire.

Because they are the main entries into a facility or a room, doorways are often prime targets for unauthorized entry into a facility or an asset. Therefore, securing doorways may be a major step in providing security at a facility.

A doorway includes four main components:

- The door, which blocks the entrance. The primary threat to the actual door is breaking or piercing through the door. Therefore, the primary security features of doors are their strength and resistance to various physical threats, such as fire or explosions.
- The door frame, which connects the door to the wall. The primary threat to a door frame is that the door can be pried away from the frame. Therefore, the primary security feature of a door frame is its resistance to prying.
- The hinges, which connect the door to the door frame. The primary threat to door hinges is that they can be removed or broken, which will allow intruders to remove the entire door. Therefore, security hinges are designed to be resistant to breaking. They may also be designed to minimize the threat of removal from the door.
- The lock, which connects the door to the door frame. Use of the lock is controlled through various security features, such as keys, combinations, etc., such that only authorized personnel can open the lock and go through the door. Locks may also incorporate other security features, such as software or other systems to track overall use of the door, to track individuals using the door, and so on.

Each of these components is integral in providing security for a doorway, and upgrading the security of only one of these components while leaving the other components unprotected may not increase the overall security of the doorway. For example, many facilities upgrade door locks as a basic step in increasing their security. However, if the facilities do not also focus on increasing security for the door hinges or the door frame, the door may remain vulnerable to being removed from its frame, thereby defeating the increased security of the door lock.

The primary attribute for the security of a door is its strength. Many security doors are four- to 20-gauge hollow metal doors consisting of steel plates over a hollow cavity reinforced with steel stiffeners to give the door extra stiffness and rigidity. This increases resistance to blunt force used to try to penetrate through the door. The space between the stiffeners may be filled with specialized materials to provide fire-, blast-, or bullet resistance to the door.

The Windows and Doors Manufacturers Association has developed a series of performance attributes for doors. These include

- Structural resistance
- Forced entry resistance
- Hinge-style screw resistance
- Split resistance

- Hinge resistance
- Security rating
- Fire resistance
- Bullet resistance
- Blast resistance

The first five bullets provide information on a door's resistance to standard physical breaking and prying attacks. These tests are used to evaluate the strength of the door and the resistance of the hinges and the frame in a standardized way. For example, the Rack Load Test simulates a prying attack on a corner of the door. A test panel is restrained at one end, and a third corner is supported. Loads are applied and measured at the fourth corner. The Door Impact Test simulates a battering attack on a door and frame using impacts of two hundred foot-pounds by a steel pendulum. The door must remain fully operable after the test. It should be noted that door glazing is also rated for resistance to shattering, and other damage. Manufacturers will be able to provide security ratings for these features of a door as well.

Door frames are an integral part of doorway security because they anchor the door to the wall. Door frames are typically constructed from wood or steel, and they are installed such that they extend for several inches over the doorway that has been cut into the wall. For added security, frames can be designed to have varying degrees of overlap with, or wrapping over, the underlying wall. This can make prying the frame from the wall more difficult. A frame formed from a continuous piece of metal (as opposed to a frame constructed from individual metal pieces) will prevent prying between pieces of the frame.

Many security doors can be retrofit into existing frames; however, many security door installations include replacing the door frame as well as the door itself. For example, bullet-resistance per Underwriter's Laboratory (UL) 752 requires resistance of the door and frame assembly, and thus replacing the door only would not meet UL 752 requirements.

Valve Lockout Devices

Valves are utilized as control elements in process piping networks. They regulate the flow of both liquids and gases by opening, closing, or obstructing a flow passageway. Valves are typically located where flow control is necessary. They can be located in-line or at pipeline tank entrance and exit points. They can serve multiple purposes in a process pipe network, including

- Redirecting and throttling flow
- Preventing backflow

- Shutting off flow to a pipeline or tank (for isolation purposes)
- Releasing pressure
- Draining extraneous liquid from pipelines or tanks
- Introducing chemicals into the process network
- As access points for sampling process water

Valves may be located either aboveground or belowground. It is critical to provide protection against valve tampering. For example, tampering with a pressure relief valve could result in a pressure buildup and potential explosion in the piping network. On a larger scale, addition of a contaminant or noncompatible chemical substance to the chemical processing system through an unprotected valve could result in the catastrophic release of that contaminant to the general population.

Different security products are available to protect aboveground versus belowground valves. For example, valve lockout devices can be purchased to protect valves and valve controls located aboveground. Vaults containing underground valves can be locked to prevent access to these valves.

As described above, a lockout device can be used as a security measure to prevent unauthorized access to aboveground valves located within petrochemical processing systems. Valve lockout devices are locks that are specially designed to fit over valves and valve handles to control their ability to be turned or seated. These devices can be used to lock the valve into the desired position. Once the valve is locked, it cannot be turned unless the locking device is unlocked or removed by an authorized individual.

Various valve lockout options are available for industrial use, including

- Cable lockouts
- Padlocked chains/cables
- Valve-specific lockouts

Many of these lockout devices are not specifically designed for use in energy sector industry (i.e., chains, padlocks) but are available from a local hardware store or manufacturer specializing in safety equipment. Other lockout devices (for example, valve-specific lockouts or valve box locks) are more specialized and must be purchased from safety or valve-related equipment vendors.

The three most common types of valves for which lockout devices are available are gate, ball, and butterfly valves. Each is described in more detail below.

- Gate valve lockouts—Gate valve lockouts are designed to fit over the operating hand wheel of the gate valve to prevent it from being turned. The lockout is secured in place with a padlock. Two types of gate valve lockouts are available: diameter-specific and adjustable. Diameter-specific lockouts are available for handles ranging from one inch to thirteen inches in diameter. Adjustable gate valve lockouts can be adjusted to fit any handle, ranging from one inch to six-plus inches in diameter.

- Ball valve lockouts—There are several different configurations available to lockout ball valves, all of which are designed to prevent rotation of the valve handle. The three major configurations available are a wedge shape for one-inch-to three inch valves, a lockout that completely covers a 3/8-inch to an 8-inch ball valve handles, and a universal lockout that can be applied to quarter-turn valves of varying sizes and geometric handle dimensions. All three types of ball valve lockouts can be installed by sliding the lockout device over the ball valve handle and securing it with a padlock.

- Butterfly valve lockouts—The butterfly valve lockout functions in a similar manner to the ball valve lockout. The polypropylene lockout device is placed over the valve handle and secured with a padlock. This type of lockout has been commonly used in the bottling industry.

A major difference between valve-specific lockout devices and the padlocked chain or cable lockouts discussed earlier is that they do not need to be secured to an anchoring device in the floor or the piping system. In addition, valve-specific lockouts eliminate potential tripping or access hazards that may be caused by chains or cable lockouts applied to valves located near walkways or frequently maintained equipment.

Valve-specific lockout devices are available in a variety of colors, which can be useful in distinguishing different valves. For example, different colored lockouts can be used to distinguish the type of liquid passing through the valve (e.g., treated, untreated, potable, petrochemical), or to identify the party responsible for maintaining the lockout. Implementing a system of different-colored locks on operating valves can increase system security by reducing the likelihood of an operator inadvertently opening the wrong valve and causing a problem in the system.

Security for Vents

Vents are installed in some aboveground energy sector storage areas to allow safe venting of off-gases. The specific vent design for any given application will vary ,depending on the design of the storage vessel. However, every vent consists of an open-air connection between the storage container and the outside environment. Although these air exchange vents are an integral part of covered or underground chemical storage containers, they also represent a potential security threat. Improving vent security by making the vents tamper-resistant or by adding other security features, such as security screens or security covers, can enhance the security of the entire processing system.

Many municipalities already have specifications for vent security at their local chemical industrial assets. These specifications typically include the following requirements:

- Vent openings are to be angled down or shielded to minimize the entrance of surface and/or rainwater into the vent through the opening.
- Vent designs are to include features to exclude insects, birds, animals, and dust.
- Corrosion-resistant materials are to be used to construct the vents.

Visual Surveillance Monitoring

Visual surveillance is used to detect threats through continuous observation of important or vulnerable areas of an asset. The observations can also be recorded for later review or use (for example, in court proceedings). Visual surveillance systems can be used to monitor various parts of production, distribution, or pumping/compressing systems, including the perimeter of a facility, outlying pumping stations, or entry or access points into specific buildings. These systems are also useful in recording individuals who enter or leave a facility, thereby helping to identify unauthorized access. Images can be transmitted live to a monitoring station, where they can be monitored in real time, or they can be recorded and reviewed later. Many energy sector facilities have found that a combination of electronic surveillance and security guards provides an effective means of facility security.

Visual surveillance is provided through a closed circuit television (CCTV) system, in which the capture, transmission, and reception of an image is localized within a closed "circuit." This is different from other broadcast images, such as over-the-air television, which is broadcast over the air to any receiver within range.

At a minimum, a CCTV system consists of

- One or more cameras
- A monitor for viewing the images
- A system for transmitting the images from the camera to the monitor

Specific attributes and features of camera systems, lenses, and lighting systems are presented in table 9.11.

MONITORING DEVICES

Earlier it was pointed out that proper security preparation really comes down to a three-legged approach: delay, respond, and detect. The third leg of security, to detect, is discussed in this section.

Table 9.11.   Attributes of Camera, Lenses, and Lighting Systems

Camera Systems	
Attribute	*Discussion*
Camera type	Major factors in choosing the correct camera are the resolution of the image required and lighting of the area to be viewed.
	• Solid state (including charge coupled devices, charge priming device, charge injection device, and metal oxide substrate)—these cameras are becoming predominant in the marketplace because of their high resolution and their elimination of problems inherent in tube cameras.
	• Thermal—These cameras are designed for night vision. They require no light and use differences in temperature between objects in the field of view to produce a video image. Resolution is low compared to other cameras, and the technology is currently expensive relative to other technologies.
	• Tube—These cameras can provide high-resolution burn out and must be replaced after one or two years. In addition, tube performance can degrade over time. Finally, tube cameras are prone to burn images in the tube replacement.
Resolution (the ability to see fine details)	User must determine the amount of resolution required depending on the level of detail required for threat determination. A high-definition focus with a wide field of vision will give an optimal viewing area.
Field of vision width	Cameras are designed to cover a defined field of vision, which is usually defined in degrees. The wider the field of vision, the more area a camera will be able to monitor.
Type of image produced (color, black and white, thermal)	Color images may allow the identification of distinctive markings, while black and white images may provide sharper contrast. Thermal imaging allows the identification of heat sources (such as human beings or other living creatures) from low-light environments; however, thermal images are not effective in identifying specific individuals (i.e., for subsequent legal processes).
Pan/Tilt/Zoom (PTZ)	Panning (moving the camera in a horizontal plane), tilting (moving the camera in a vertical plane), and zooming (moving the lens to focus on objects that are at different distances from the camera) allow the camera to follow a moving object. Different systems allow these functions to be controlled manually or automatically. Factors to be considered in PTZ cameras are the degree of coverage for pan and tilt functions and the power of the zoom lens.
Lenses	
Format	Lens format determines the maximum image size to be transmitted.
Focal length	This is the distance from the lens to the center of the focus. The greater the focal length, the higher the magnification, but the narrower the field of vision.
F number	F number is the ability to gather light. Smaller F numbers may be required for outdoor applications where light cannot be controlled as easily.

Attribute	Discussion
Distance and width approximation	The distance and width approximations are used to determine the geometry of the space that can be monitored at the best resolution.
	Lighting Systems
Intensity	Light intensity must be great enough for the camera type to produce sharp images. Light can be generated from natural or artificial sources. Artificial sources can be controlled to produce the amount and distribution of light required for a given camera and lens.
Evenness	Light must be distributed evenly over the field of view so that there are no darker or shadowy areas. If there are lighter vs. darker areas, brighter areas may appear washed out (i.e., details cannot be distinguished) while no specific objects can be viewed from darker areas.
Location	Light sources must be located above the camera so that light does not shine directly into the camera.

Source: U.S. Environmental Protection Agency, *Water and Wastewater Security Product Guide*, http://cfpub.epa.gov/ safewater/watersecurity/guide (accessed April 4, 2009).

Radiation Detection Equipment

Radioactive contaminants (radionuclides) are known health hazards that emit energetic waves (radiation) that can cause both carcinogenic and noncarcinogenic health effects. Radionuclides pose unique threats to source water supplies and distribution systems because radiation emitted from radionuclides in nuclear waste systems can affect individuals through several pathways—by direct contact with, ingestion or inhalation of, or external exposure to the contaminated waste stream. While radiation can occur naturally in some cases due to the decay of some minerals, intentional and unintentional releases of man-made radionuclides into petrochemical water feed or petrochemical processing streams is also a realistic threat.

Threats to the nuclear power sector from radioactive contamination could involve the facility or its assets becoming contaminated and preventing workers from accessing and operating the facility/assets.

COMMUNICATION INTEGRATION

In this section, those devices necessary for communication and integration of nuclear power sector industrial processing operations, such as electronic controllers, two-way radios, and wireless data communications, are discussed.

With regard to security applications, electronic controllers are used to automatically activate equipment (such as lights, surveillance cameras, audible alarms, or locks) when they are triggered. Triggering could be in response to variety of scenarios, including tripping of an alarm or a motion sensor; breaking of a window or a glass door; variation in vibration sensor readings; or simply input from a timer.

Two-way wireless radios allow two or more users who have their radios tuned to the same frequency to communicate instantaneously with each other without the radios being physically lined together with wires or cables.

Wireless data communications devices are used to enable transmission of data between computer systems and/or between a SCADA server and its sensing devices, without individual components being physically linked together via wires or cables. In industrial petrochemical processing systems, these devices are often used to link remote monitoring stations (i.e., SCADA components) or portable computers (i.e., laptops) to computer networks without using physical wiring connections.

Electronic Controllers

An electronic controller is a piece of electronic equipment that receives incoming electric signals an uses preprogrammed logic to generate electronic output signals based on the incoming signals. While electronic controllers can be implemented for any application that involves inputs and outputs (for example, control of a piece of machinery in a factory), in a security application, these controllers essentially act as the system's "brain" and can respond to specific security-related inputs with preprogrammed output response. These systems combine the control of electronic circuitry with a logic function such that circuits are opened and closed (and thus equipment is turned on and off) through some preprogrammed logic. The basic principle behind the operation of an electrical controller is that it receives electronic inputs from sensors or any device generating an electrical signal (for example, electrical signals from motion sensors) and then uses its preprogrammed logic to produce electrical outputs (for example, these outputs could turn on power to a surveillance camera or to an audible alarm). Thus, these systems automatically generate a preprogrammed, logical response to a preprogrammed input scenario.

The three major types of electronic controllers are timers, electromechanical relays, and programmable logic controllers (PLCs), which are often called "digital relays." Each of these types of controller is discussed in more detail below.

Timers use internal signal/inputs (in contrast to externally generated inputs) and generate electronic output signals at certain times. More specifically, timers control electric current flow to any application to which they are connected and can turn the current on or off on a schedule prespecified by the user. Typical timer range (amount of time that can be programmed to elapse before the timer activates linked equipment) is from 0.2 seconds to 10 hours, although some of the more advanced timers have ranges of up to 60 hours. Timers are useful in fixed applications that don't require frequent schedule changes. For example, a timer can be used to turn on the lights in a room or building at a certain time every day. Timers are usually connected to their own power supply (usually 120-240 V).

In contrast to timers, which have internal triggers based on a regular schedule, electromechanical relays and PLCs have both external inputs and external outputs. However, PLCs are more flexible and more powerful than are electromechanical relays, and thus this section focuses primarily on PLCs as the predominant technology for security-related electronic control applications.

Electromechanical relays are simple devices that use a magnetic field to control a switch. Voltage applied to the relay's input coil creates a magnetic field, which attracts an internal metal switch. This causes the relay's contacts to touch, closing the switch and completing the electrical circuit. This activates any linked equipment. These types of systems are often used for high-voltage applications, such as in some automotive and other manufacturing processes.

Two-Way Radios

Two way radios, as discussed here, are limited to a direct unit-to-unit radio communication, either via single unit-to-unit transmission and reception or via multiple handheld units to a base station radio contact and distribution system. Radio frequency spectrum limitations apply to all handheld units and are directed by the FCC. This also distinguishes a handheld unit from a base station or base station unit, such as those used by an amateur (ham) radio operator, which operate under different wave length parameters.

Two-way radios allow a user to contact another user or group of users instantly on the same frequency and to transmit voice or data without the need for wires. They use "half-duplex" communications, or communication that can be only transmitted or received; it cannot transmit and receive simultaneously. In other words, only one person may talk, while other personnel with radio(s) can only listen. To talk, the user depresses the talk·button and speaks into the radio. The audio then transmits the voice wirelessly to the receiving radios. When the speaker has finished speaking and the channel has cleared, users on any of the receiving radios can transmit, either to answer the first transmission or to begin a new conversation. In addition to carrying voice data, many types of wireless radios also allow the transmission of digital data, and these radios may be interfaced with computer networks that can use or track these data. For example, some two-way radios can send information such as global positioning system (GPS) data or the ID of the radio. Some two-way radios can also send data through a SCADA system.

Wireless radios broadcast these voice or data communications over the airwaves from the transmitter to the receiver. While this can be an advantage in that the signal emanates in all directions and does not need a direct physical connection to be received at the receiver, it can also make the communications vulnerable to being blocked, intercepted, or otherwise altered. However, security features are available to ensure that the communications are not tampered with.

Wireless Data Communications

A wireless data communication system consists of two components: a "wireless access point" (WAP), and a "wireless network interface card" (sometimes also referred to as a "client"), which work together to complete the communications link. These wireless systems can link electronic devices, computers, and computer systems together using radio waves, thus eliminating the need for these individual components to be directly connected together through physical wires. While wireless data communications have widespread application in water and wastewater systems, they also have limitations. First, wireless data connections are limited by the distance between components (radio waves scatter over a long distance and cannot be received efficiently unless special directional antennas are used). Second, these devices only function if the individual components are in a direct line of sight with each other, since radio waves are affected by interference from physical obstructions. However, in some cases, repeater units can be used to amplify and retransmit wireless signals to circumvent these problems. The two components of wireless devices are discussed in more detail below.

(1) WAP—The WAP provides the wireless data communication service. It usually consists of a housing (which is constructed from plastic or metal, depending on the environment it will be used in) containing a circuit board, flash memory that holds software, one of two external ports to connect to existing wired networks; a wireless radio transmitter/receiver; and one or more antenna connections. Typically, the WAP requires a one-time user configuration to allow the device to interact with the local area network (LAN). This configuration is usually done via a web-driven software application accessed via a computer.

(2) Wireless Network Interface Card/Client—A wireless card is a piece of hardware that is plugged into a computer and enables that computer to make a wireless network connection. The card consists of a transmitter, functional circuitry, and a receiver for the wireless signal, all of which work together to enable communication among the computer, its wireless transmitter/receiver, and its antenna connection. Wireless cards are installed in a computer through a variety of connections, including USB adapters, laptop card bus (PCMCIA), or desktop peripheral (PCI) cards. As with the WAP, software is loaded onto the user's computer, allowing configuration of the card so that it may operate over the wireless network

Two of the primary applications for wireless data communications systems are to enable mobile or remote connections to a LAN and to establish wireless communications links between SCADA remote telemetry units (RTUs) and sensors in the field.

Wireless car connections are usually used for LAN access from mobile computers. Wireless cards can also be incorporated into RTUs to allow them to communicate with sensing devices that are located remotely.

CYBER PROTECTION DEVICES

Various cyber protection devices are currently available for use in protecting nuclear power sector computer systems. These protection devices include anti-virus and pest eradication software, firewalls, and network intrusion hardware/software. These products are discussed in this section.

Anti-Virus and Pest Eradication Software

Anti-virus programs are designed to detect, delay, and respond to programs or pieces of code that are specifically designed to harm computers. These programs are known as "malware." Malware can include computer viruses, worms, and Trojan horse programs (programs that appear to be benign but which have hidden harmful effects).

Pest eradication tools are designed to detect, delay, and respond to "spyware" (strategies that websites use to track user behavior, such as by sending "cookies" to the user's computer), and hacker tools that track keystrokes (keystroke loggers) or passwords (password crackers).

Viruses and pests can enter a computer system through the Internet or through infected floppy discs or CDs. They can also be placed onto a system by insiders. Some of these programs, such as viruses and worms, move within a computer's drives and files, or between computers if the computers are networked to each other. This malware can deliberately damage files, utilize memory and network capacity, crash application programs, and initiate transmissions of sensitive information from a PC. While the specific mechanisms of these programs differ, they can infect files, even the basic operating program of the computer firmware/hardware.

The most important features of an anti-virus program are its abilities to identify potential malware and to alert a user before infection occurs, as well as its ability to respond to a virus already resident on a system. Most of these programs provide a log so that the user can see what viruses have been detected and where they were detected. After detecting a virus, the anti-virus software may delete the virus automatically, or it may prompt the user to delete the virus. Some programs will also fix files or programs damaged by the virus.

Various sources of information are available to inform the general public and computer system operators about new viruses being detected. Since anti-virus programs use signatures (or snippets of code or data) to detect the presence of a virus, periodic updates are required to identify new threats. Many anti-virus software providers offer free upgrades that are able to detect and respond to the latest viruses.

Firewalls

A firewall is an electronic barrier designed to keep computer hackers, intruders, or insiders from accessing specific data files and information on an energy sector's computer network or other electronic/computer systems. Firewalls operate by evaluating and then filtering information coming through a public network (such as the Internet) into the utility's computer or other electronic system. This evaluation can include identifying the source or destination addresses and ports and allowing or denying access based on this identification.

There are two methods used by firewalls to limit access to the utility's computers or other electronic systems from the public network:

- The firewall may deny all traffic unless it meets certain criteria.
- The firewall may allow all traffic through unless it meets certain criteria.

A simple example of the first method is to screen requests to ensure that they come from an acceptable (i.e., previously identified) domain name and Internet protocol address. Firewalls may also use more complex rules that analyze the application data to determine whether the traffic should be allowed through. For example, the firewall may require user authentication (i.e., use of a password) to access the system. How a firewall determines what traffic to let through depends on which network layer it operates at and how it is configured. Some of the pros and cons of various methods to control traffic flowing in and out of the network are provided in table 9.12.

Firewalls may be a piece of hardware, a software program, or an appliance card that contains both.

Advanced features that can be incorporated into firewalls allow for the tracking of attempts to log onto the local area network system. For example, a report of successful and unsuccessful log-in attempts may be generated for the computer specialist to analyze. For systems with mobile users, firewalls allow remote access into the private network by the use of secure log-on procedures and authentication certificates. Most firewalls have a graphical user interface for managing the firewall.

In addition, new Ethernet firewall cards that fit in the slot of an individual computer bundle additional layers of defense (like encryption and permit/deny) for individual computer transmissions to the network interface function. These new cards have only a slightly higher cost than traditional network interface cards.

Network Intrusion Hardware/Software

Network intrusion detection and prevention system are software- and hardware-based programs designed to detect unauthorized attacks on a computer network system.

Table 9.12. Pros and Con of Various Firewall Methods for Controlling Network Access

Method	Description	Pros	Cons
Packet filtering	Incoming and outgoing packets (small chunks of data) are analyzed against a set of filters. Packets that make it through the filters are sent to the requesting system and all others are discarded. There are two types of packet filtering: static (the most common) and dynamic.	Static filtering is relatively inexpensive, and little maintenance is required. It is well-suited for closed environments where access to or from multiple addresses is not allowed.	Leaves permanent open holes in the network; allows direct connection to internal hosts by external sources; offers no user authentication, method can be un-manageable in large networks.
Proxy service	Information from the Internet is retrieved by the firewall and then sent to the requesting system and vice versa. In this way, the firewall can limit the information made known to the requesting system, making vulnerabilities less apparent.	Only allows temporary open holes in the network perimeter. Can be used for all types of internal protocol services.	Allows direct connections to internal hosts by external clients; offers no user authentication
Stateful pattern recognition	This method examines and compares the contents of certain key parts of an information packet against a database of acceptable information. Information traveling from inside the firewall to the outside is monitored for specific defining characteristics, then incoming information is compared to these characteristics. If the comparison yield a reasonable match, the information is allowed through. If not, the information is discarded.	Provides a limited time window to allow packets of information to be sent; does not allow any direct connections between internal and external hosts; supports user-level authentication.	Slower than packet filtering; does not support all types of connections.

Source: U.S. Environmental Protection Agency, *Water and Wastewater Security Product Guide,* http://cfpub.epa.gov/safewater/watersecurity/guide (accessed April 4, 2009).

While other applications, such as firewalls and anti-virus software, share similar objectives with network intrusion systems, network intrusion systems provide a deeper layer of protection beyond the capabilities of these other systems because they evaluate patterns of computer activity rather than specific files.

It is worth noting that attacks may come from either outside or within the system (i.e., from an insider), and that network intrusion detection systems may be more applicable for detecting patterns of suspicious activity from inside a facility (i.e., accessing sensitive data and so on.) than are other information technology solutions.

Network intrusion detection systems employ a variety of mechanisms to evaluate potential threats. The types of search and detection mechanisms are dependent upon the level of sophistication of the system. Some of the available detection methods include

- Protocol analysis—Protocol analysis is the process of capturing, decoding, and interpreting electronic traffic. The protocol analysis method of network intrusion detection involves the analysis of data captured during transactions between two or more systems or devices, and the evaluation of these data to identify unusual activity and potential problems. Once a problem is isolated and recorded, problems or potential threats can be linked to pieces of hardware or software. Sophisticated protocol analysis will also provide statistics and trend information on the captured traffic.
- Traffic anomaly detection—Traffic anomaly detection identifies potential threatening activity by comparing incoming traffic to "normal" traffic patterns, and identifying deviations. It does this by comparing user characteristics against thresholds and triggers defined by the network administrator. This method is designed to detect attacks that span a number of connections rather than a single session.
- Network honeypot—This method establishes nonexistent services in order to identify potential hackers. A network honeypot impersonates services that don't exist by sending fake information to people scanning the network. It identifies attackers when they attempt to connect to the service. There is no reason for legitimate traffic to access these resources because they don't exist; therefore, any attempt to access them constitutes an attack.
- Anti-intrusion detection system evasion techniques—These methods are designed for attackers who may be trying to evade intrusion detection system scanning. They include methods called IP defragmentation, TCP streams reassembly, and deobfuscation.

While these detection systems are automated, they can only indicate patterns of activity, and a computer administrator or other experienced individual must interpret activities to determine whether or not they are potentially harmful. Monitoring the logs generated by these systems can be time consuming, and there may be a learning curve to determine a baseline of "normal" traffic patterns from which to distinguish potential suspicious activity.

REFERENCES AND RECOMMENDED READING

Garcia, M. L. 2001. *The design and evaluation of physical protection systems.* Boston: Butterworth-Heinemann.

IBWA. 2004. *Bottled water safety and security.* Alexandria, VA: International Bottled Water Association.

NAERC. 2002. *Security guidelines for the electricity sector.* Washington, DC: North American Electric Reliability Council.

NEI. 2009. *Key issues: Nuclear power plant security.* http://www.nei.org/keyissues/safetyandsecurity/plantsecurity (accessed December 27, 2009).

Schneier, B. 2000. *Secrets & lies.* New York: Wiley.

U.S. EPA. 2005. *Water and wastewater security product guide.* http://cfpub.epa.gov.safewater/watersecurity/guide (accessed April 4, 2009).

10

Emergency Response

When Mayor Rudi Giuliani made the above statement to Police Commissioner Bernard Kerik at the World Trade Center site, September 11, 2001, to a point and to a degree, one of the first (and not to be forgotten) gross understatements of the twenty-first century had been uttered. Indeed, for citizens of the United States of America, the 9/11 events raised our level of consciousness, awareness, fear, and questions of what to do next in "uncharted territory." Actually, when you get right down to it, 9/11 generated more questions than anything else. Many are still asking the following questions today:

Why?

Why would anyone have the audacity to attack the United States?

What kind of cold-blooded killers would even think of conducting such an event?

Who were those Islamic radicals who perpetrated 9/11?

What were the terrorists' goal(s)?

Why do they lack any respect for human life . . . including their own?

Why?

Why weren't we prepared or ready for such an attack?

Why had we not foreseen such an event?

Why were our emergency responders so undermanned, ill prepared, and ill equipped to handle such a disaster?

What took the military fighter planes so long to respond?

What did our government really know (if anything) before the events occurred?

Could anyone have prevented it?

Bottom line questions: Why us? Hell, why anyone?

Why?

These and several other questions continue to resonate today, especially with the occurrence of subsequent terrorist attacks such as the infamous shoe bomber and underwear bomber attempts to blow up jet airplanes filled with innocent people. No doubt terrorism and terrorists will continue to haunt and attempt to destroy us for some time to come.

Maybe we ask post-9/11-related questions because of who we are, what we are, and what we are not. That is, because we are Americans we are free, uninhibited thinkers who think what we say and say what we think—isn't America great! Most Americans are softhearted and sympathetic to those in need—compassion is the very nature and soul of being American. Americans are not born terrorists; they are not born into a terrorist regime; they are not raised with fear in their hearts—they are not afraid every time they leave their homes and go about their daily business—are they? Suicide bombers and other like terrorists are those who occupy some other faraway place, definitely not America, and they are definitely not American. Right?

Notwithstanding exceptions to the rule, such as Timothy McVeigh (a so-called red-blooded American, war-decorated hero, born and raised in America) and Bruce Edwards Ivins, who mailed the anthrax, terrorism was foreign to us.

Today, from a safety and security point of view, based on the events of 9/11 and the anthrax events, we should no longer be asking why. Instead, we should not waste our time, money, and energy asking why or in pointing a finger of blame at our government, military, 9/11 emergency responders, and/or the terrorists. We should stop asking why and shift our mindset to asking "what if." The point is we need to accept the fact that there are folks out there who do not share our view of the American way of life. Earlier, with regard to security preparedness, we pointed to the need to ask what-if questions. Simply, what-if analysis is a proactive approach used to prevent or mitigate certain disasters or extreme events—those human- or nature-generated. Obviously, asking and properly answering what-if questions has little effect on preventing the actions of Mother Nature, such as earthquakes, tornadoes, hurricanes (Katrina-type events), and others. On the other hand, it is true that what-if questions, when properly posed and answered (with results), can reduce the death toll and overall damage caused by these natural disasters. We are certainly aware that these natural events are possible, probable, and likely, and their effects can be horrendous—beyond tragic. The irony is apparent, however, especially when we ask how many of us are actually willing to move away from or out of earthquake zones, hurricane and tornado allies, and floodplains to live somewhere else?

The fact is we do not possess a crystal ball to foretell the future. Thus, what-if questions prepare us to react and respond to certain contingencies. And respond we must, because there are certain events we simply can't prevent. The best response to an event we can't prevent is summed up by the Boy Scouts Motto: Be Prepared!

NUCLEAR POWER PLANT CONTINGENCY PLANNING

The NRC (2009) reexamined the role of emergency preparedness (EP) for protection of the public near nuclear power plants following the accident at the Three Mile Island nuclear power plant in 1979. The accident showed the need for improved planning, response, and communication by federal, state, and local government to prevent reactor accidents. Although the NRC remained vigilant over the years, the events of 9/11 prompted a new focus on emergency preparedness and a further review of the threat environment. The NRC now considers new threat scenarios and protections in emergency preparedness in light of the threat of terrorist attacks.

Nuclear power plant owners, government agencies, and state and local officials, as well as thousands of volunteers and first responders, have worked together for more than twenty years to create a system of emergency preparedness and response that will serve the public well in the unlikely event of an emergency. The nuclear power plants' emergency plans include preparations for evacuation, sheltering, or other actions to protect plant operators and the residents near nuclear power plants in the event of a serious incident.

Since commercial nuclear power plants began operating in the United States, there have been no physical injuries or fatalities from exposure to radiation from the plants among members of the U.S. public. Even the country's worst nuclear power plant accident at Three Mile Island resulting in no identifiable heath impacts.

FEDERAL OVERSIGHT

In the U.S., 104 commercial nuclear power reactors are licensed to operate at sixty-five sites in thirty-one states. For each site, there are onsite and offsite emergency plans to assure that adequate protective measures can be taken to protect the public in the event of a radiological emergency.

Federal oversight of emergency preparedness for licensed nuclear power plants is shared by the NRC and Federal Emergency Management Agency (FEMA). This sharing is facilitated through a Memorandum of Understanding (MOU). The MOU is responsive to the president's decision of December 7, 1979, that FEMA take the lead in overseeing off-site planning and response, and that NRC assist FEMA in carrying out this role. The NRC has statutory responsibility for the radiological health and safety of the public by overseeing on-site preparedness and has overall authority for both on-site and off-site emergency preparedness.

Before a plant is licensed to operate, the NRC must have "reasonable assurance that adequate protective measures can and will be taken in the event of a radiological emergency." The NRC's decision of reasonable assurance is based on licensees complying with NRC regulations and guidance. In addition, licensees and area response organizations must demonstrate they can effectively implement emergency plans and procedures during period evaluated exercises. As part of the Reactor Oversight Process, the NRC reviews licensees' emergency planning procedures and training. These reviews include regular drills and exercises that assist licensees in identifying areas for improvement, such as in the interface of security operations and emergency preparedness. Each plant owner is required to exercise its emergency plan with the NRC, FEMA, and offsite emergency plans. Licensees also self-test their emergency plans regularly by conducting drills. Each plant's performance in drills and exercises can be accessed through the NRC website at this address: http://www.nrc.gov/NRR/OVERSIGHT/ASSESS/index.html.

FEMA takes the lead in initially reviewing and assessing the offsite planning and response and in assisting state and local governments, while the NRC reviews and assesses the on-site planning and response. FEMA findings and determinations as to the adequacy and capability of implementing off-site plans are communicated to the NRC. The NRC reviews the FEMA findings and determinations as well as the on-site findings. The NRC then makes a determination on the overall state of emergency preparedness. These overall findings and determinations are used by the NRC to make radiological health and safety decisions before the issuing of licenses and in the continuing oversight of operating reactors. The NRC has the authority to take action, including shutting down any reactor deemed not to provide reasonable assurance of the protection of public health and safety.

EMERGENCY PLANNING ZONES

For planning purposes, the NRC defines two emergency planning zones (EPZs) around each nuclear power plant. The exact size and configuration of the zones vary from plant to plant due to local emergency response needs and capabilities, popula-tion, land characteristics, access routes, and jurisdictional boundaries. The two types of EPZs are

- The plume exposure pathway EPZ extends approximately ten miles in radius around a plant. Its primary concern is the exposure of the public to, and the inhalation of, airborne radioactive contamination.
- The ingestion pathway EPZ extends approximately fifty miles in radius around a plant. Its primary concern is the ingestion of food and liquid that is contaminated by radioactivity.

EMERGENCY CLASSIFICATION

Emergency classification is a set of plant conditions that indicate a level of risk to the public. Nuclear power plants use the four emergency classifications listed below in order of increasing severity:

- **Notification of Unusual Event**—Under this category, events are in process or have occurred that indicate potential degradation in the level of safety of the plant. No release of radioactive material requiring off-site response or monitoring is expected unless further degradation occurs.
- **Alert**—If an alert is declared, events are in process or have occurred that involve an actual or potential substantial degradation in the level of safety of the plant. Any releases of radioactive material from the plant are expected to be limited to a small fraction of the Environmental Protection agency (EPA) protective action guides (PAGs). Additional information regarding PAGs can be found on the EPA website at: http://ww.epa.gov/radiation/rert/pags.html.
- **Site Area Emergency**—A site area emergency involves events in process or which have occurred that result in actual or likely major failures of plant functions needed for protection of the public. Any releases of radioactive material are not expected to exceed the EPA PAGs except near the site boundary.
- **General Emergency**—A general emergency involves actual or imminent substantial core damage or melting of reactor fuel with the potential for loss of containment in-tegrity. Radioactive releases during a general emergency can reasonably be expected to exceed the EPA PAGs for more than the immediate site area.

PROTECTIVE ACTIONS

The NRC's regulations are designed to mitigate accident consequences and minimize radiation exposure to the public through protective actions. When a radiological emergency occurs, nuclear power plant personnel evaluate plant conditions and make protective action recommendations to the state and local government agencies on how to protect the population. Based on the recommendation and independent assessment of other local factors, the state or local government agencies are responsible for making decisions on the actions necessary to protect the public and for relaying these decisions to the public.

Factors that affect protective action decisions include plant conditions, competing events, weather, evacuation times, shelter factors, how quickly an incident develops, how short-lived a release of radiation may be, and other conditions.

EVACUATION, SHELTERING, AND THE USE OF POTASSIUM IODIDE

Protective actions considered for a radiological emergency include evacuation, sheltering, and, as a supplement to these, the prophylactic use of potassium iodide (KI), as appropriate. Under most conditions, evacuation may be preferred to remove the public from further exposure to radioactive material. However, under some conditions, people may be instructed to take shelter in their homes, schools, or office buildings. Depending on the type of structure, sheltering can significantly reduce a person's dose compared to remaining outside. In certain situations, KI is used as supplement to sheltering.

Evacuation does not always call for completely emptying the ten-mile zone around a nuclear power plant. In most cases, the release of radioactive material from a plant during a major incident would move with the wind, not in all directions surrounding the plant. The release should also expand and become less concentrated as it travels away from a plant. Therefore, evacuations should be mapped to anticipate the path of the release. Generally, as a minimum, in the event of a general emergency, a two-mile ring around the plant is evacuated, along with people living in the five-mile zone directly downwind and slightly to either side of the projected path of the release. This "keyhole" pattern helps account for potential wind shifts and fluctuations in the release pattern. Evacuation beyond five miles is assessed as the accident progresses. Also in response to a general emergency, people living in the remainder of the ten-mile zone will most likely be advised to go indoors to monitor Emergency Alert System broadcasts.

Sheltering is a protective action that keeps people indoors, such as at home, the office, school, or a shopping mall to reduce exposure to radioactive material. It may be appropriate to shelter when the release of radioactive material is known to be short-term or controlled by the nuclear power plant operator. Additional information on evacuation and sheltering can be found on the NRC website at http://www.nrc.gov/what-we-do/emerg-prepredness/evacuation-sheltering.html.

Another protective action in the ten-mile EPA involves KI, a compound that helps prevent the thyroid from absorbing radioactive iodine, one of several radioactive materials that could be present in a release from a nuclear power plant accident. If taken within the appropriate time and at the appropriate dosage, KI blocks the radioactive iodine from being absorbed by the thyroid gland and reduces the risk of thyroid cancers and other diseases. KI does not protect against any other inhaled radioactive materials, nor will it offer protection from external exposure to radiation. The Food and Drug Administration (FDA) has determined that KI is a safe and effective drug when used for this purpose. However, there may be risks and potential side effects in using KI, including gastrointestinal disturbances, allergic reactions, and iodide goiter and hypothyroidism.

TERRORISM AND EMERGENCY PREPAREDNESS

After September 2001, the NRC examined how terrorist-based events might challenge existing emergency preparedness. The NRC's formal evaluation determined that, in view of the threat environment, the emergency preparedness planning basis remained valid. While a terrorist event might alter the initial response to an event, the consequences of the event will be the same whether it was caused by terrorism or a safety accident.

The nuclear power reactor's emergency plans are periodically updated and are designed to be flexible to identify, evaluate, and react to the wide spectrum of emergency conditions. The NRC recognized how the terrorism threat affects emergency planning when it issued orders and guidance to nuclear power plants after September 2001. These orders and guidance include interim measures dealing with how increased security affects implementation of emergency plans. Nuclear industry groups and federal, state, and local government agencies assisted in the prompt implementation of these measures and participated in drills and exercises to test these new planning elements. The NRC has reviewed licensees' commitments to address these requirements and verified the implementation through inspections to ensure public health and safety.

REFERENCES AND RECOMMENDED READING

Brauer, R. L.. 1994. *Safety and health for engineers*. New York: Van Nostrand Reinhold.

CoVan, J. 1995. *Safety engineering*. New York: John Wiley & Sons, 1995.

FEMA. 1981. *Planning guide and checklist for hazardous materials contingency plans* (FEMA-10). Washington, DC: Federal Emergency Management Agency.

Healy, R. J. 1969. *Emergency and disaster planning*. New York: Wiley.

NRC. 2009. *Backgrounder on emergency preparedness at nuclear power plants*, available at http://www.nrc.gov/reading-rm/doc-collections/fact-sheets/emerg-plan-prep-nuc -power-bg.html (accessed December 30, 2009).

Office of the Federal Register. 1987. 29 CFR 1910.120. Washington, DC: Office of the Federal Register.

Smith, A. J. 1980. *Managing hazardous substances accidents.* New York: McGraw-Hill.

Spellman, F. R. 1997. *A guide to compliance for process safety management planning* (PSM/RMP). Lancaster, PA: Technomic.

U.S. Army Corps of Engineers. 1987. *Safety and health requirements manual,* revised edition (EM 385-1-1). Washington, DC: U.S. Army Corps of Engineers.

U.S. DOE. 2008. *Emergency support function #12—Energy annex.* Washington, DC: U.S. Department of Energy.

U.S. EPA. 2002. *Water utility response, recovery & remediation guidance for man-made and/or technological emergencies.* Washington, DC: U.S. Environmental Protection Agency.

U.S. EPA. 2003. *Large water system emergency response plan outline: Guidance to assist community water systems in complying with the Public Health Security and Bioterrorism Preparedness and Response Act of 2002* (EPA 810-F-03-007). Washington, DC: U.S. Environmental Protection Agency, www.epa.gov/safewater/security (accessed June 2006).

Epilogue

The Paradigm Shift

The 9/11 shift: There is a new worldview in the making.

The events of 9/11 dramatically changed this nation and focused us on combating terrorism. As a result, in 2003 and subsequent years, the Department of Homeland Security (DHS), in conjunction with members from the general public, state and local agencies, and private groups concerned with the safety of critical infrastructures, established a Security Working Group (WSWG) to consider and make recommendations on infrastructure security issues. For example, the WSWG identified active and effective security practices for critical infrastructure and provided an approach for adopting these practices. It also recommended mechanisms to provide incentives that facilitate broad and receptive response among critical infrastructure sectors to implement active and effective security practices. Finally, WSWG recommended mechanisms to measure progress and achievements in implementing active and effective security practices and to identify barriers to implementation.

The WSWG recommendations on security are structured to maximize benefits to critical industries by emphasizing actions that have the potential both to improve the quality or reliability of service and to enhance security. These recommendations, based on original recommendations from the 2003 National Drinking Water Advisor Council (NDWAC), were designed primarily, as the name suggests, for use by water systems of all types and sizes, including systems that serve fewer than 3,300 people. However, it is the authors' opinion, based on personal experience, that NDWAC's recommendations, when properly adapted to applicable circumstances, can be applied to any and all critical infrastructure sectors, including the nuclear power sector.

The NDWAC identified fourteen features of active and effective security programs that are important to increasing security and relevant across the broad range of utility circumstances and operating conditions. U.S. EPA (2003) points out that the fourteen features are, in many cases, consistent with the steps needed to maintain technical,

265

management, and operational performance capacity related to overall water quality; as mentioned, these steps can be applied to other critical infrastructures as well. Even though nuclear power plants are heavily regulated by NRC and other regulatory agencies/requirements, the "fourteen feature" template provided below can augment the regulated security requirements. Moreover, many facilities may be able to adopt some of the features with minimal, if any, capital investment. The point is we simply cannot overprepare for any potential contingency; this is especially the case in the era of shoe- and underwear-bomber-terrorists.

FOURTEEN FEATURES OF ACTIVE AND EFFECTIVE SECURITY

It is important to point out that the fourteen features of active and effective programs emphasize that "one size does not fit all" and that there will be variability in security approaches and tactics among critical infrastructure facilities, based on industry-specific circumstances and operating conditions. The fourteen features

- Are sufficiently flexible to apply to all nuclear industries, regardless of size
- Incorporate the idea that active and effective security programs should have measurable goals and time lines
- Allow flexibility for nuclear power sector facilities to develop specific security approaches and tactics that are appropriate to industry-specific circumstances

Nuclear power sector facilities can differ in many ways, including

- Transportation supply source (rail, air, water, pipeline, lines, or ground)
- Number of supply sources
- Energy processing capacity
- Operation risk
- Location risk
- Security budget
- Spending priorities
- Political and public support
- Legal barriers
- Public versus private ownership

Nuclear power sector industrial facilities should address security in an informed and systematic way, regardless of these differences. Nuclear power sector facilities need to fully understand the specific, local circumstances and conditions under which they operate and develop a security program tailored to those conditions. The goal in identifying common features of active and effective security programs is to achieve consistency in security program outcomes among critical infrastructure sector industrial facilities, while allowing for and encouraging facilities to develop utility-specific security approaches and tactics. The features are based on a comprehensive "security management layering system" approach that incorporates a combination of public involvement and awareness, partnerships, and physical, chemical, operational, and design controls to increase overall program performance. They address industry security in four functional categories: *organization, operation, infrastructure,* and *external.* These functional categories are discussed in greater detail below.

- **Organizational**—There is always something that can be done to improve security. Even when resources are limited, the simple act of increasing organizational attentiveness to security may reduce vulnerability and increase responsiveness. Preparedness itself can help deter attacks. The first step to achieving preparedness is to make security a part of the organizational culture, so that it is in the day-to-day thinking of frontline employees, emergency responders, and management of every energy sector facility in this country. To successfully incorporate security into "business as usual," there must be a strong commitment to security by organization leadership and by the supervising body, such as the board of stockholders. The following features address how a security culture can be incorporated into an organization.
- **Operational**—In addition to having a strong culture and awareness of security within an organization, an active and effective security program makes security part of operational activities, from daily operations, such as monitoring of physical access controls, to scheduled annual reassessments. Nuclear power sector industries will often find that by implementing security into operations they can also reap cost benefits and improve the quality or reliability of the energy service.
- **Infrastructure**—These recommendations advise utilities to address security in all elements of energy sector industry infrastructure—from source to distribution and through processing and product delivery.
- **External**—Strong relationships with response partners and the public strengthen security and public confidence. Two of the recommended features of active end effective security programs address this need.

Fourteen Features

Feature 1. Make an explicit and visible commitment of the senior leadership to security. Nuclear power sector industrial facilities should create an explicit, easily communicated, enterprise-wide commitment to security, which can be done through

- Incorporating security into a utility-wide mission or vision statement, addressing the full scope of an active and effective security program—that is, protection of worker/public health, worker/public safety, and public confidence, and that is part of core day-to-day operations.

- Developing an enterprise-wide security policy or set of policies.

Nuclear power sector industries should use the process of making a commitment to security as an opportunity to raise awareness of security throughout the organization, making the commitment visible to all employees and customers, and to help every facet of the enterprise recognize the contribution it can make to enhancing security.

Feature 2. Promote security awareness throughout the organization.
The objective of a security culture should be to make security awareness a normal, accepted, and routine part of day-to-day operations. Examples of tangible efforts include

- Conducting employee training
- Incorporating security into job descriptions
- Establishing performance standards and evaluations for security
- Creating and maintaining a security tip line and suggestion box for employees
- Making security a routine part of staff meetings and organization planning
- Creating a security policy

Feature 3. Assess vulnerabilities and periodically review and update vulnerability assessments to reflect changes in potential threats and vulnerabilities.
Because circumstances change, nuclear power sector industrial facilities should maintain their understanding and assessment of vulnerabilities as a "living document" and continually adjust their security enhancement and maintenance priorities. Nuclear power sector industrial facilities should consider their individual circumstances and establish and implement a schedule for review of their vulnerabilities.

Assessments should take place once every three to five years at a minimum. Nuclear power sector industries may be well served by doing assessments annually.

The basic elements of sound vulnerability assessments are:

- Characterization of the chemical processing system, including its mission and objectives
- Identification and prioritization of adverse consequences to avoid
- Determination of critical assets that might be subject to malevolent acts that could result in undesired consequences
- Assessment of the likelihood (qualitative probability) of such malevolent acts from adversaries
- Evaluation of existing countermeasures

- Analysis of current risk and development of a prioritized plan for risk reduction

Feature 4. Identify security priorities and, on an annual basis, identify the resources dedicated to security programs and planned security improvements, if any.
Dedicated resources are important to ensure a sustained focus on security. Investment in security should be reasonable, considering utilities' specific circumstances. In some circumstances, investment may be as simple as increasing the amount of time and attention that executives and managers give to security. Where threat potential or potential consequences are greater, greater investment likely is warranted.

This feature establishes the expectation that nuclear power facilities should, through their annual capital, operations, and maintenance and through staff resources plans, identify and set aside resources consistent with their specific identified security needs. Security priorities should be clearly documented and should be reviewed with utility executives at least once per year as part of the traditional budgeting process.

Feature 5. Identify managers and employees who are responsible for security, and establish security expectations for all staff.

- Explicit identification of security responsibilities is important for development of a security culture with accountability.
- At minimum, energy sector industrial facilities should identify a single, designated individual responsible for overall security, even if other security roles and responsibilities will likely be dispersed throughout the organization.
- The number and depth of security-related roles will depend on a utility's specific circumstances.

Feature 6. Establish physical and procedural controls to restrict access to nuclear power industrial infrastructure to only those conducting authorized, official business and to detect unauthorized physical intrusions.
Examples of physical access controls include fencing critical areas, locking gates and doors, and installing barriers at site access points. Monitoring for physical intrusion can include maintaining well-lighted facility perimeters, installing motion detectors, and utilizing intrusion alarms. The use of neighborhood watches, regular employee rounds, and arrangements with local police and fire departments can support identifying unusual activity in the vicinity of facilities.

Examples of procedural access controls include inventorying keys, changing access codes regularly, and requiring security passes to pass gates and access sensitive areas. In addition, utilities should establish the means to readily identify all employees, including contractors and temporary workers with unescorted access to facilities.

Feature 7. Employee protocols for detection of contamination consistent with the recognized limitations in current contaminant detection, monitoring, and surveillance technology.

Until progress can be made in development of practical and affordable online contaminant monitoring and surveillance systems, most nuclear power sector industrial facilities must use other approaches to contaminant monitoring and surveillance.

Many utilities already measure the above parameters (and many others) on a regular basis to control plant operations and confirm chemical mixture quality; more closely monitoring these parameters may create operational benefits for facilities that extend far beyond security, such as reducing operating costs and chemical usage. Nuclear power sector industrial facilities also should thoughtfully monitor customer complaints and improve connections with local public health networks to detect public health anomalies. Customer complaints and public health anomalies are an important way to detect potential contamination problems and other environmental quality concerns.

Feature 8. Define security-sensitive information; establish physical, electronic, and procedural controls to restrict access to security-sensitive information; detect unauthorized access; and ensure information and communications systems will function during emergency response and recovery.

Protecting IT systems largely involves using physical hardening and procedural steps to limit the number of individuals with authorized access and to prevent access by unauthorized individuals. Examples of physical steps to harden SCADA and IT networks include installing and maintaining firewalls and screening the network for viruses. Examples of procedural steps include restricting remote access to data networks and safeguarding critical data through backups and storage in safe places. Nuclear power facilities should strive for continuous operation of IT and telecommunications systems, even in the event of an attack, by providing uninterruptible power supply and backup systems, such as satellite phones.

In addition to protecting IT systems, security-sensitive information should be identified and restricted to the appropriate personnel. Security-sensitive information could be contained within

- Facility maps and blueprints
- Operations details
- Hazardous material utilization and storage
- Tactical level security program details
- Any other information on utility operations or technical details that could aid in planning or executing an attack

Identification of security-sensitive information should consider all ways that utilities might use and make public information (e.g., many chemical industrial facilities may at times engage in competitive bidding processes for construction of new facilities or infrastructure). Finally, information critical to the continuity of day-to-day operations should be identified and backed up.

Feature 9. Incorporate security considerations into decisions about acquisition, repair, major maintenance, and replacement of physical infrastructure; include consideration of opportunities to reduce risk through physical hardening and adoption of inherently lower-risk design and technology options.

Prevention is a key aspect of enhancing security. Consequently, consideration of security issues should begin as early as possible in facility construction (i.e., it should be a factor in building plans and designs). However, to incorporate security considerations into design choices, chemical facilities need information about the types of security design approaches and equipment that are available and the performance of these designs and equipment in multiple dimensions. For example, nuclear power sector facilities would want to evaluate not only the way that a particular design might contribute to security, but also how that design would affect the efficiency of day-to-day plant operations and worker safety.

Feature 10. Monitor available threat-level information and escalate security procedures in response to relevant threats.

Monitoring threat information should be a regular part of a security program manager's job, and utility-, facility-, and region-specific threat levels and information should be shared with those responsible for security. As part of security planning, nuclear power sector facilities should develop systems to access threat information, procedures that will be followed in the event of increased industry or facility threat levels, and should be prepared to put these procedures in place immediately, so that adjustments are seamless. Involving local law enforcement and FBI is critical.

Nuclear power sector facilities should investigate what networks and information sources might be available to them locally and at the state and regional level. If a utility cannot gain access to some information networks, attempts should be made to align with those who can and will provide effective information to the energy sector facility.

Feature 11. Incorporate security considerations into emergency response and recovery plans, test and review plans regularly, and update plans to reflect changes in potential threats, physical infrastructure, chemical processing operations, critical interdependencies, and response protocols in partner organizations.

Nuclear power sector facilities should maintain response and recovery plans as "living documents." In incorporating security considerations into their emergency response and recovery plans, nuclear power facilities also should be aware of the

National Incident Management System (NIMS) guidelines, established by DHS, and of regional and local incident management commands and systems, which tend to flow from the national guidelines.

Nuclear power sector facilities should consider their individual circumstances and establish, develop, and implement a schedule for review of emergency response and recovery plans. Nuclear power sector facility plans should be thoroughly coordinated with emergency response and recovery planning in the larger community. As part of this coordination, a mutual aid program should be established to arrange in advance for exchanging resources (personnel or physical assets) among agencies within a region in the event of an emergency or disaster that disrupts operation. Typically, the exchange of resource is based on a written formal mutual aid agreement. For example, Florida's Water-Wastewater Agency Response Network (FlaWARN), deployed after Hurricane Katrina and allowed the new "utilities helping utilities" network to respond to urgent requests from Mississippi for help to bring facilities back online after the hurricane.

The emergency response and recovery plans should be reviewed and updated as needed annually. This feature also establishes the expectation that chemical facilities should test or exercise their emergency response and recovery plans regularly.

Feature 12. Develop and implement strategies for regular, ongoing security-related communications with employees, response organizations, rate-setting organizations, and customers.

An active and effective security program should address protection of public health, public safety (including infrastructure), and public confidence. Energy sector facilities should create an awareness of security and an understanding of the rationale for their overall security management approach in the communities they reside in and/or serve.

Effective communication strategies consider key messages; who is best equipped/ trusted to deliver the key messages; the need for message consistency, particularly during an emergency; and the best mechanisms for delivering messages and for receiving information and feedback from key partners. The key audiences for communication strategies are utility employees, response organizations, and customers.

Feature 13. Forge reliable and collaborative partnerships with the communities served, managers of critical interdependent infrastructure, response organizations, and other local utilities.

Effective partnerships build collaborative working relationships and clearly define roles and responsibilities so that people can work together seamlessly if an emergency should occur. It is important for nuclear power sector facilities within a region and neighboring regions to collaborate and establish a mutual aid program with neighboring utilities, response organizations, and sectors, such as the power sector, on which utilities rely or that they impact. Mutual aid agreements provide for help from other organizations that is prearranged and can be accessed quickly and efficiently in the event of a terrorist attack or natural disaster. Developing reliable and collaborative partnerships involves reaching out to managers and key staff and other organizations to build reciprocal understanding and to share information about the facility's security concerns and planning. Such efforts will maximize the efficiency and effectiveness of a mutual aid program during an emergency response effort, as the organizations will be familiar with one another's circumstances and thus will be better able to serve one another.

It is also important for nuclear power sector facilities to develop partnerships with the communities and customers they serve. Partnerships help build credibility within communities and establish public confidence in utility operations. People who live near nuclear power sector facility structures can be the eyes and ears of the facility and can be encouraged to notice and report changes in operating procedures or other suspicious behaviors.

Nuclear power sector facilities and public health organizations should establish formal agreements on coordination to ensure regular exchange of information between facilities and public health organizations, and outline roles and responsibilities during response to and recovery from an emergency. Coordination is important at all levels of the public health community—national public health, county health agencies, and health-care providers, such as hospitals.

Feature 14. Develop nuclear power facility–specific measures of security activities and achievements, and self-assess against these measures to understand and document program progress.
Although security approaches and tactics will be different depending on nuclear power utility-specific circumstances and operating conditions, we recommend that all energy sector facilities monitor and measure a number of common types of activities and achievements, including existence of program policies and procedures, training, testing, and implementing schedules and plans.

The Fourteen-Feature Matrix

In the following, a matrix of recommended measures to assess effectiveness of a nuclear power sector facility's security program is presented. Each feature is grouped according to its functional category: organization, operation, infrastructure, and external.

Ultimately, the goal of implementing the fourteen security features (and all other security provisions) is to create a significant improvement in nuclear power sector industry on a national scale by reducing vulnerabilities and, therefore, risk to public health from terrorist attacks and natural disasters. To create a sustainable effect, the energy sector as a whole must not only adopt and actively practice the features, but also incorporate them into "business as usual."

Table Epilogue 1.   Fourteen Features of Active and Effective Security Matrix

Features	Checklist: Potential Measures of Progress
Organizational Features	
Feature 1—Explicit commitment to security	Does a written, enterprise-wide security policy exist, and is the policy reviewed regularly and updated as needed?
Feature 2—Promote security awareness	Are incidents reported in a timely way, and are lessons learned from incident responses reviewed and, as appropriate, incorporated into future utility security efforts?
Feature 5—Defined security roles and employee expectations	Are managers and employees who are responsible for security identified?
Operational Features	
Feature 3—Vulnerability assessment up to date	Are reassessments of vulnerabilities made after incidents, and are lessons learned and other relevant information incorporated into security practices?
Feature 4—Security resources and implementation priorities	Are security priorities clearly identified, and to what extent do security priorities have resources assigned to them?
Feature 7—Contamination detection	Is there a protocol/procedure in place to identify and respond to suspected contamination events?
Feature 10—Threat-level-based protocols	Is there a protocol/procedure of responses that will be made if threat levels change?
Feature 11—Emergency response plan tested and up to date	Do exercises address the full range of threats—physical, cyber, and contamination—and is there a protocol/procedure to incorporate lessons learned from exercises and actual response into updates to emergency response and recovery plans?
Feature 14—Industry-specific measures and self-assessment	Does the facility perform self-assessment a least annually?
Infrastructure Features	
Feature 6—Intrusion detection and access control	To what extent are methods to control access to sensitive assets in place?
Feature 8—Information protection and continuity	Is there a procedure to identify and control security-sensitive information, is information correctly categorized, and how do control measures perform under testing?
Feature 9—Design and construction standards	Are security considerations incorporated into internal utility design and construction standards for new facilities/infrastructure and major maintenance projects?
External Features	
Feature 12—Communications	Is there a mechanism for utility employees, partners, and the community to notify the utility of suspicious occurrences and other security concerns?
Feature 13—Partnerships	Have reliable and collaborative partnerships with customers, managers of independent interrelated infrastructure, and response organizations been established?

Source: U.S. EPA (2003), Active and effective water security programs, http://cfpub.epa.gov/safewater/watersecurity/14 features.cfm (accessed June 2006).

REFERENCES AND RECOMMENDED READING

U.S. EPA. 2003. *Active and effective water security programs.* http://cfpub.epa.gov/safewater/watersecurity/14 features.cfm (accessed June 2006).

Index

About the Authors

Frank R. Spellman is Assistant Professor of Environmental Health at Old Dominion University. He is a professional member of the American Society of Safety Engineers, the Water Environment Federation, and the Institute of Hazardous Materials Managers. He is also a Board Certified Safety Professional and Board Certified Hazardous Materials Manager with more than 35 years of experience in environmental science and engineering. He is the author of more than seventy books, including *Water Infrastructure Protection and Homeland Security* (GI, 2007) and *Food Supply Protection and Homeland Security* (GI, 2008).

Melissa L. Stoudt graduated summa cum-laude from Old Dominion University in 2008, with a Bachelors of Science in Environmental Health. While at the University Melissa was involved in many projects—most notable was research for the Centers for Pediatric Research in Norfolk, Virginia, involving particulate matter in a public housing community. She went on to present and win awards for this research at several conferences across the country. Melissa is also a veteran of the United States Navy from which she was honorably discharged in 2002. Melissa currently works as a radiological controls training instructor at an atomic power laboratory.

Breinigsville, PA USA
15 December 2010
251521BV00003B/1/P